高 等 学 校 教 材

化学专业英语

赵修毅　林志强　主编

化学工业出版社
·北京·

内 容 简 介

《化学专业英语》精选了四大化学（无机化学、有机化学、分析化学、物理化学）的基本知识、相关知识的短篇阅读文章及化学实验报告与化学期刊文章的写作方法。另外还附有元素的英文名、常用化学专业术语、常用实验室装置及无机物定性分析方法。

《化学专业英语》可作为高等院校化学及相关专业的专业英语教学用书，也可作为从事化学及相关专业的科技工作者的参考用书。

图书在版编目（CIP）数据

化学专业英语/赵修毅，林志强主编．—北京：化学工业出版社，2021.10（2023.9重印）
高等学校教材
ISBN 978-7-122-39649-5

Ⅰ.①化⋯ Ⅱ.①赵⋯②林⋯ Ⅲ.①化学-英语-高等学校-教材 Ⅳ.①O6

中国版本图书馆CIP数据核字（2021）第152565号

责任编辑：马泽林　杜进祥　　　　　　装帧设计：关　飞
责任校对：宋　玮

出版发行：化学工业出版社（北京市东城区青年湖南街13号　邮政编码100011）
印　　装：三河市双峰印刷装订有限公司
787mm×1092mm　1/16　印张13¾　字数350千字　2023年9月北京第1版第4次印刷

购书咨询：010-64518888　　　　　　　售后服务：010-64518899
网　　址：http://www.cip.com.cn
凡购买本书，如有缺损质量问题，本社销售中心负责调换。

定　价：39.00元　　　　　　　　　　　　　　　　　　　版权所有　违者必究

前言

《化学专业英语》是大学化学及相关专业本科生的一门重要课程。其目的在于提升学生的化学英文文献的阅读能力及英语沟通交流能力，为学生毕业后从事与化学相关的教学与科研等工作打下扎实的专业英语基础。

本书参考国外化学教材的内容及编写方式，先介绍化学专业各二级学科的概况，之后再进行该学科的重要概念介绍，并使用短篇阅读文章加强读者对化学原理的认识及应用。本书还简要介绍了化学实验报告及化学期刊文章的写作方法，帮助学生有效地解决化学专业英语使用的实际问题。

本书分为7章，包含化学主要领域及基本知识的介绍（第1~5章）、生活中与化学相关的短文（第6章）及实验报告和化学文章的写作指引（第7章）。此外，元素的英文名、常用化学专业术语、常用实验室装置及无机物定性分析方法在附录中列出。在第1~6章的每个章节后面列出了词汇、注解与问题，便于读者阅读、学习与思考。书中的重要单词或短语做加粗处理，更加便于读者抓住重点。书中标斜体的内容为某些单词的定义、英文字的前/后缀或希腊文/拉丁文。

本书由中山大学赵修毅和林志强主编，由赵修毅提出写作的框架和编写方式，并做最后统稿及定稿工作。赵修毅编写第1~6章及附录A~C，林志强编写第7章、附录D和进行了各章的修改工作。此外，张丽和杨建成做了查找资料、整理及校核工作。

本书在编写的过程中，得到了中山大学本科教学质量工程项目（教材建设项目）的资助，得到了中山大学和化学学院领导的大力支持与热情帮助，在此表示衷心的感谢。

由于编者水平有限，书中疏漏在所难免，恳请同行专家及读者批评指正。

编者
2021年5月于广州

Contents

Chapter 1 General Chemistry / 001

1.1 Chemistry: The Central Science ········· 001
1.2 The Scientific Method ········· 002
1.3 Matter: Physical State and Chemical Composition ········· 004
1.4 Elements and the Periodic Table ········· 009
1.5 Chemical Reactions in Aqueous Solution ········· 012

Chapter 2 Inorganic Chemistry / 016

2.1 Nomenclature of Inorganic Compounds ········· 016
2.2 Nomenclature of Coordination Compounds ········· 021
2.3 Ionic and Covalent Bonds ········· 023
2.4 Coordination Compounds ········· 030
2.5 The Halogens ········· 034

Chapter 3 Organic Chemistry / 041

3.1 What is Organic Chemistry? ········· 041
3.2 Classes of Organic Compounds ········· 043
3.3 Nomenclature of Organic Compounds ········· 045
3.4 Hydrocarbons ········· 053
3.5 Nucleophilic Substitution Reactions ········· 065
3.6 Isomerism ········· 070
3.7 Organic Polymers ········· 074

Chapter 4   Analytical Chemistry   / 081

4.1 Introduction to Analytical Chemistry ········· 081
4.2 Spectrochemical Methods ········· 084
4.3 Chromatography ········· 091
4.4 Thin-Layer Chromatography ········· 095
4.5 NMR Spectroscopy ········· 100
4.6 Mass Spectrometry ········· 104

Chapter 5   Physical Chemistry   / 109

5.1 Introduction to Physical Chemistry ········· 109
5.2 Chemical Equilibrium ········· 114
5.3 Thermodynamics ········· 119
5.4 Solutions ········· 124
5.5 Catalysis ········· 130
5.6 Electrochemical Cells ········· 134

Chapter 6   Chemistry in Life   / 141

6.1   Nitroglycerine ········· 141
6.2   Coupling of Reactions ········· 142
6.3   Haber Process ········· 144
6.4   Air Bags ········· 146
6.5   Self-Cleaning Windows ········· 147
6.6   Lithium-Ion Batteries ········· 148
6.7   Liquid Crystals ········· 150
6.8   Superconductors ········· 152
6.9   The Discovery of Prodrugs ········· 154
6.10 Magnetic Resonance Imaging ········· 156

Chapter 7   Writing Lab Reports and Papers   / 158

7.1 Some Guidelines for Writing Lab Reports ········· 158

7.2 Some Guidelines for Writing Scientific Papers ·· 168

Appendices / 173

Appendix A Names of Chemical Elements ··· 173
Appendix B Common Chemical Terms ·· 176
Appendix C Some Laboratory Apparatus ·· 208
Appendix D Simple Guidelines to Qualitative Analysis of Inorganic
            Compounds ··· 210

References / 214

Chapter 1

General Chemistry

1.1 Chemistry: The Central Science

Chemistry is sometimes referred to as "the central science" due to its interconnection with a vast array of other scientific disciplines, such as technology, engineering, and mathematics. Chemistry plays a vital role in biology, medicine, materials science, forensics, environmental science, and many other fields (Figure 1.1). The fundamental principles of physics are essential for understanding many aspects of chemistry, and there is extensive overlap between many subdisciplines within the two fields, such as nuclear chemistry and chemical physics. Mathematics, computer science, and information theory provide powerful tools that help us describe, calculate, interpret, and understand the world of chemistry. Biochemistry is developed from biology and chemistry. It helps us understand many complex factors and processes that keep living organisms alive. Chemical engineering, nanotechnology, forensics and materials science combine empirical findings and chemical principles to generate useful substances, ranging from petroleum to fabrics to microelectronics. Food science, brewing, and agricultural chemistry keep us away from starvation. Medicine, pharmacology, botany, and biotechnology identify and produce useful substances that help keep us healthy. Environmental science, oceanography, atmospheric science, geology, astronomy and cosmology integrate many chemical concepts to help us protect our living world and better understand the universe.

What chemical changes are important to our daily life? Digesting food and absorption of nutrients, synthesizing polymers that are used to make clothing, containers, cookware, and tires, and refining crude oil into gasoline and other useful compounds are just a few examples. Actually, chemistry is the study of the composition and properties of matter and the changes it undergoes.

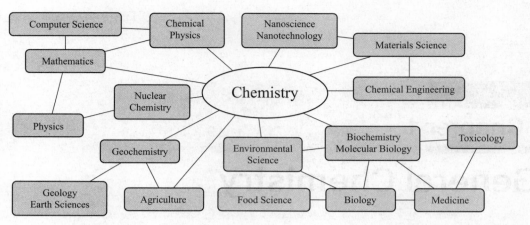

Figure 1.1 Chemistry is essential for a good understanding of a wide range of scientific disciplines

New Words and Expressions

astronomy /əˈstrɒnəmi/ n. 天文学
biotechnology /ˌbaɪəʊtekˈnɒlədʒi/ n. 生物技术
cosmology /kɒzˈmɒlədʒi/ n. 宇宙学
crude oil n. 原油
forensics /fəˈrensɪks/ n. 法医学

nanotechnology /ˌnænəʊtekˈnɒlədʒi/ n. 纳米技术
oceanography /ˌəʊʃəˈnɒɡrəfi/ n. 海洋学
pharmacology /ˌfɑːməˈkɒlədʒi/ n. 药物学
discipline /ˈdɪsəplɪn/ n. 学科
veterinary science n. 兽医学

Notes

bio- 表示"生命，生物"，如 biotechnology（生物技术），biology（生物学），biochemistry（生物化学），biography（传记）。

Questions

Why is chemistry often called the central science?

1.2 The Scientific Method

Although two scientists rarely approach the same problem in exactly the same way, they use guidelines for the practice of science known as the **scientific method**. Figure 1.2 is a flow chart of the scientific method. We begin by collecting information, or data, by observation and experiment. Our ultimate purpose, however, is not collecting data but rather finding a pattern or sense of order in our observations and understanding the origin of this order. As we collect more data, we may discover some patterns that lead us to a rough explanation, or **hypothesis**, that helps us in designing further experiments. An important feature of a good hypothesis is that it proposes an underlying mechanism that explains our observations and can be used to make some predictions about new experiments. If a hypothesis is sufficiently general and repeatedly successful in predicting results of future experiments, it is called a theory. A **theory** is *an explanation of the general causes of certain phenomena, with considerable evidence or facts to support it*. For example, Einstein's theory of relativity was revolutionary in a sense that space and time are not independent. Note that it was more than just a hypothesis because it could be used to make effective predictions that could be tested experimentally. The results of these experiments were generally in good agreement with Einstein's predictions and were not be explained by earlier theories satisfactorily.

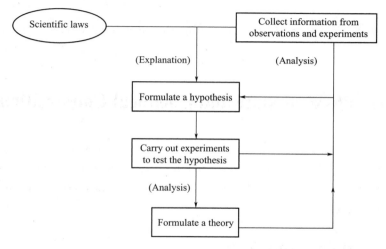

Figure 1.2 A flow chart of the scientific method

In spite of the landmark achievements of Einstein's theory, scientists can never say the theory is "proven". A theory that has excellent predictive power today may not work as well in the future as more data and improved scientific equipment are developed. Thus, science is always a work in progress.

Eventually, we may be able to tie together a great number of observations in a **scientific law**, which is *a concise verbal statement or mathematical equation that summarizes a broad variety of observations and experiences*. We tend to think of scientific laws as the basic rules under which nature operates. However, it is not so much that matter obeys these laws, but rather that these laws describe the behavior of matter. As we proceed through this text, we will rarely have the opportunity to discuss the doubts, conflicts, clashes of personalities, and revolutions of perception that have led to our present scientific ideas. You need to be aware that just because we can spell out the results of science so concisely and neatly in textbooks it does not mean scientific progress is smooth, certain, and predictable. Some of the ideas we learned from the text books need centuries to develop and involves many scientists. We gain our view of the natural world by standing on the shoulders of the scientists who came before us. Take advantage of this view. As you study, exercise your imagination. Don't be afraid to ask daring questions when they occur to you. You may be fascinated by what you discover!

New Words and Expressions

hypothesis /haɪˈpɒθəsɪs/ n. 假说　　　　　theory /ˈθɪərɪ/ n. 理论

Questions

1. Is it possible to know how many experiments are required to verify a natural law? Explain.
2. If you wish to test a theory, describe the necessary characteristics of a suitable experiment.

1.3　Matter: Physical State and Chemical Composition

Matter is anything that has mass and occupies space. In general, we can classify matter by its physical state as a solid, liquid, or gas, or by its chemical composition as an element, compound, or mixture.

1.3.1　Solid, Liquids, and Gases

Commonly, a given kind of matter exists in different physical forms under different conditions. For example, water exists as ice (solid), liquid water, and steam (gaseous water). When a substance goes from one state of matter to another, the process is called a change of state, or **phase change**. Solids tend to maintain their shapes when subjected to outside forces. Liquids and gases are **fluids** which can flow and change their shapes easily.

What distinguishes a gas from a liquid is the characteristic of compressibility (and its opposite, expansibility). A gas is easily compressible, whereas a liquid is not. You can put more and more air into a tire, which increases only slightly in volume. In fact, a given quantity of gas can fill a container of almost any size. A smaller quantity would expand to fill the container. A larger quantity could be compressed to fill the same space. By contrast, if you were to try to force more liquid water into a closed glass bottle that was already full of water, it would burst.

These two characteristics, rigidity (or fluidity) and compressibility (or expansibility), can be used to frame definitions of the three common **states of matter**.

Solid: the form of matter characterized by rigidity; a solid is relatively incompressible and has a definite shape and occupies a definite volume.

Liquid: the form of matter that is a relatively incompressible fluid; a liquid has a fixed volume but no fixed shape.

Gas: the form of matter that is an easily compressible fluid; a given quantity of gas will fit into a container of almost any size and shape.

The term **vapor** is often used to refer to the gaseous state of any kind of matter that normally exists as a liquid or a solid.

1.3.2 Elements, Compounds, and Mixtures

To understand how matter is classified by its chemical composition, we must first distinguish between physical and chemical changes and between physical and chemical properties. A **physical change** is *a change in the form of matter but not in its chemical identity*. Changes of physical state are examples of physical changes. The process of dissolving one material in another is a further example of a physical change. For instance, you can dissolve sodium chloride in water. The result is a clear solution, like pure water, though many of its other characteristics are different from those of pure water. The water and sodium chloride in this solution retain their chemical identities and can be separated by some method that depends on physical changes.

Distillation is a common way to separate the sodium chloride and water components of this solution. You place the colorless solution in a flask to which a water condenser is attached. The solution in the flask is heated to boiling (Boiling involves the formation of bubbles of the water in the body of the solution). Steam escapes from the surface of the solution and passes into the water condenser, where the vapor changes back to liquid water. The liquid water is collected in another flask, called a **receiver**. When all of the water evaporates off, the original flask now contains the solid sodium chloride. Thus, by means of physical changes (the change of liquid water to vapor and back to liquid), you have separated the sodium chloride and water that you had earlier mixed together.

A **chemical change**, or **chemical reaction**, is *a change in which one or more kinds of matter are transformed into a new kind of matter or several new kinds of matter*. During the rusting of iron, iron combines with oxygen in the air to form a new compound, which is commonly called rust. This process is a chemical change. Iron and oxygen combine chemically and cannot be separated by any physical methods. To recover the iron and oxygen from rust, a chemical change or a series of chemical changes is required.

We characterize or identify a material by its various properties, which may be either physical or chemical. A **physical property** is *a characteristic that can be observed for a material without changing its chemical identity*. Examples are physical state (solid, liquid, or gas), melting point, color and electrical conductivity. Physical properties can be extensive or intensive. Extensive properties, such as mass and volume, depend on the amount of matter present. While intensive properties, such as color and density, do not depend on the amount of matter present. For example, a large chunk of silver is the same color as a small chunk of silver. Sometimes, intensive properties can be used to identify a substance. For example, quartz and diamond look very alike, but they have different density. On the other hand, a **chemical property** is *a characteristic of a material involving its chemical change*. A chemical property of sodium is its ability to react with water to produce hydrogen gas and sodium hydroxide solution.

The various materials around us are either substances or mixtures of substances. A **substance** is *a kind of matter that cannot be separated into other kinds of matter by any physical process*. Earlier you saw that when sodium chloride is dissolved in water, it is possible to separate the sodium chloride from the water by the physical process of distillation. However, sodium chloride is itself a substance and cannot be separated by physical process into new materials. Similarly, pure water is a substance.

No matter what its source, a substance always has the same characteristic properties. For example, no matter how sodium chloride is obtained by burning sodium in chlorine or from seawater, it is a white solid melting at 801℃.

Chemists have characterized millions of substances that are composed of a very small number of elements. Lavoisier was the first to define an **element** as *a substance that cannot be decomposed by any chemical reaction into simpler substances*. In 1789, Lavoisier listed 33 substances as elements, of which more than 20 are still so regarded. Today 118 elements are known.

Most substances are compounds. *Two or more elements combine chemically to form a new substance is called a* **compound**. By the end of the eighteenth century, Lavoisier and others had carefully examined many compounds and experimentally showed that all of them were made of the elements in definite proportions by mass. Joseph Louis Proust (1754—1826) established the **law of definite proportions** (also known as the **law of constant composition**): *a pure compound, whether its source, always contains definite or constant proportions of the elements by mass*. For example, 1.0000 g of potassium chloride always contains 0.5244 g of potassium and 0.4756g of chlorine, chemically combined. Potassium chloride has definite proportions of potassium and chlorine. It means that potassium chloride has a definite or constant composition.

Most of the materials around us are mixtures. A **mixture** is *a material that can be separated by physical methods into two or more substances*. Unlike a pure compound, a mixture has variable composition. Its composition depends on the relative amount of substances you mixed. Each component of the mixture retains its own set of physical and chemical characteristics. For example, sodium chloride can be separated from its aqueous solution by simple distillation.

Mixtures are classified into two types. A **heterogeneous mixture** consists of *two or more physically distinct components, each with different properties*. An example of a heterogeneous mixture is iron filings and sand. In this case, iron can be separated from sand with a magnet. A **homogeneous mixture** (also known as a solution) is *a mixture that is uniform in its* *composition throughout the mixture.* When sugar is dissolved in water, you obtain a homogeneous mixture, or a sugar solution. Air is a gaseous solution, mainly containing nitrogen, oxygen, rare gases, carbon dioxide and water vapor, which are physically mixed but not chemically combined together.

A **phase** is *one of several different homogeneous materials present in the portion of matter under study.* A heterogeneous mixture of salt and sugar is said to be composed of two different phases: one of the phases is salt; the other is sugar. Similarly, ice cubes in water are said to be composed of two phases: one phase is ice; the other is liquid water. Ice floating in an aqueous solution of sodium chloride also consists of two phases, ice and the liquid solution. Note that a phase may be either a pure substance in a particular state or a solution in a particular state (solid, liquid, or gaseous). Also, the portion of matter under consideration may consist of several phases of the same substance.

Figure 1.3 summarizes the relationships among element, compounds, and mixtures. Matter is either substances or mixtures. Substances can be mixed by physical processes, and other physical processes can be used to separate the mixtures into substances. Substances are either elements or compounds. Elements may react chemically to yield compounds, and compounds may be decomposed by chemical reactions into elements.

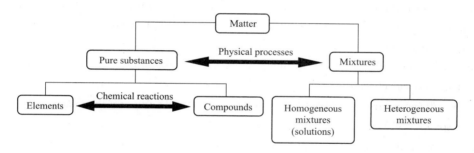

Figure 1.3 Relationships among elements, compounds, substances and mixtures

New Words and Expressions

chemical reaction n. 化学反应
compound /ˈkɒmpaʊnd/ n. 化合物
condenser /kənˈdensə(r)/ n. 冷凝器
distillation /ˌdɪstɪˈleɪʃn/ n. 蒸馏
element /ˈelɪmənt/ n. 元素
flask /flɑːsk/ n. 烧瓶
fluid /ˈfluːɪd/ n. 流体
heterogeneous mixture n. 非均相混合物；多相混合物
homogeneous mixture n. 均相混合物
matter /ˈmætə(r)/ n. 物质
mixture /ˈmɪkstʃə(r)/ n. 混合物
phase /feɪz/ n. 相
receiver /rɪˈsiːvə(r)/ n. 接收器
substance /ˈsʌbstəns/ n. 物质

Notes

1. hetero- 表示"异类的",如 heterogeneity(多相性),heterogeneous(多相的),heterocyclic(杂环的)。

2. homo- 表示"同类的",如 homogeneous(均相的),homocentric(同中心的),homochromous(单色的)。

3. A chemical change, or chemical reaction, is a change in which one or more kinds of matter are transformed into a new kind of matter or several new kinds of matter.

参考译文:化学变化或化学反应是一种或多种物质转变成一种或几种新物质的变化。

4. A physical property is a characteristic that can be observed for a material without changing its chemical identity.

参考译文:物理性质是物质不需要经过化学变化就表现出来的性质。

Questions

1. Solid iodine, contaminated with sand, was heated until the iodine vaporized. The violet vapor was then deposited on the cold surface of a condenser to yield the pure solid iodine. Is it a physical change or a chemical change?

2. Solid iodine and zinc metal powder were mixed and ignited to give a white powder ZnI_2. Is it a physical change or a chemical change?

3. The following are properties of some substances, decide whether each is a physical or a chemical property: (a) Pure water freezes at 0℃. (b) Calcium carbonate fizzes when immersed in vinegar. (c) Lithium is a soft and silvery metal. (d) Potassium bromide dissolves in water. (e) Hydrogen sulfide smells like rotten eggs.

4. Which of the followings are pure substances and which are mixtures? For each, list all of the different phases present: (a) Bromine liquid and its vapor. (b) Baking powder containing sodium hydrogen carbonate and potassium hydrogen tartrate.

5. Consider the following separation of materials. State whether a physical process or a chemical reaction is involved in each separation: (a) Sugar crystals are separated from a sugar syrup by evaporation of water. (b) Gold is obtained from river sand by panning (allowing the heavy metal to settle in flowing water). (c) Mercury and oxygen are obtained by heating mercury (Ⅱ) oxide.

6. Indicate whether each of the following materials is a pure substance, a heterogeneous mixture, or a solution: (a) Seawater. (b) Bromine. (c) Milk. (d) Gasoline.

1.4 Elements and the Periodic Table

1.4.1 Names and Symbols of Chemical Elements

All atoms with the same number of protons are atoms of the same element and they have the same atomic number, Z. Now 118 elements have been known. Each element has a name and distinctive symbol. The names of elements and their symbols come from many sources. Some names from Latin, Greek, or German words describing a characteristic property of the element. Others are named for the country or location of their discovery or to honor famous scientist. For example, the name for the element helium is derived from the Greek word *helios*, which means sun. The name Germanium comes from *Germania*, the Latin name for Germany. Element number 101 was named Mendelevium in honor of Russian chemist Dmitri Mendeleev. The symbols of elements as well as their names (English and Chinese) are listed in Appendix A.

1.4.2 The Periodic Table

The periodic table consists of 118 elements (Figure 1.4). More than half of the elements in the periodic table were discovered in the nineteenth century. According to the similar physical and chemical properties, chemists arranged the known elements in vertical columns called groups of families, leading to the development of the modern periodic table.

The elements are arranged in **periods**, horizontal rows, in order of increasing atomic number (the number of protons in the nucleus of an atom). The first period contains just two elements, hydrogen (H) and helium (He). The second and third periods each contain eight elements: lithium (Li) through neon (Ne), and sodium (Na) through argon (Ar), respectively. The fourth and fifth periods each contain 18 elements: potassium (K) through krypton (Kr), and rubidium (Rb) through xenon (Xe), respectively. The sixth period is a long period with 32 elements. To fit this period to a table that is held to a maxima width of 18 elements, we extract 14 elements of the period and place them at the bottom of the table. This series of 14 elements follows lanthanum ($Z=57$), and these elements are called the **lanthanides**. In the seventh period, 14 elements are also extracted from this period. Because the elements in this series follow actinium ($Z=89$), they are called the **actinides**.

Most elements can be classified as metals and nonmetals. A **metal** is a good conductor of heat and electricity, whereas a **nonmetal** is usually a poor conductor of heat and electricity. A **metalloid** is an element with properties that are intermediate between those metals and nonmetals. Only seventeen elements are nonmetals and eight elements are metalloids. The remaining elements in the periodic table are metals.

A vertical column of elements in the periodic table is known as a **group**. Groups of

1 H hydrogen																	2 He helium
3 Li lithium	4 Be beryllium											5 B boron	6 C carbon	7 N nitrogen	8 O oxygen	9 F fluorine	10 Ne neon
11 Na sodium	12 Mg magnesium											13 Al aluminum	14 Si silicon	15 P phosphorus	16 S sulfur	17 Cl chlorine	18 Ar argon
19 K potassium	20 Ca calcium	21 Sc scandium	22 Ti titanium	23 V vanadium	24 Cr chromium	25 Mn manganese	26 Fe iron	27 Co cobalt	28 Ni nickel	29 Cu copper	30 Zn zinc	31 Ga gallium	32 Ge germanium	33 As arsenic	34 Se selenium	35 Br bromine	36 Kr krypton
37 Rb rubidium	38 Sr strontium	39 Y yttrium	40 Zr zirconium	41 Nb niobium	42 Mo molybdenum	43 Tc technetium	44 Ru ruthenium	45 Rh rhodium	46 Pd palladium	47 Ag silver	48 Cd cadmium	49 In Indium	50 Sn tin	51 Sb antimony	52 Te tellurium	53 I iodine	54 Xe xeon
55 Cs cesium	56 Ba barium	57 La lathanum	72 Hf hafnium	73 Ta tantalum	74 W tungsten	75 Re rhenium	76 Os osmium	77 Ir iridium	78 Pt platinum	79 Au gold	80 Hg mercury	81 Tl thallium	82 Pb lead	83 Bi bismuth	84 Po polonium	85 At astatine	86 Rn radon
87 Fr francium	88 Ra radium	89 Ac actinium	104 Rf rutherfordium	105 Db dubnium	106 Sg seaborgium	107 Bh bohrium	108 Hs hassium	109 Mt meitnerium	110 Ds darmstadium	111 Rg roentgenium	112 Cn copernicium	113 Nh nihonium	114 Fl flerovium	115 Mc moscovium	116 Lv livermorium	117 Ts tennessine	118 Og oganesson

58 Ce cerium	59 Pr praseodymium	60 Nd neodymium	61 Pm promethium	62 Sm samarium	63 Eu europium	64 Gd gadolinium	65 Tb terbium	66 Dy dysprosium	67 Ho holmium	68 Er erbium	69 Tm thulium	70 Yb ytterbium	71 Lu lutetium
90 Th thorium	91 Pa protactinium	92 U uranium	93 Np neptunium	94 Pu plutonium	95 Am americium	96 Cm curium	97 Bk berkelium	98 Cf californium	99 Es einsteinium	100 Fm fermium	101 Md mendelevium	102 No nobelium	103 Lr lawrencium

Figure 1.4 The periodic table of the elements

elements sometimes are referred to collectively by their group number (Group ⅠA, Group ⅡA, and so on). Group ⅠA elements (Li, Na, K, Rb, Cs, and Fr) are called the **alkali metals**, and the Group ⅡA elements (Be, Mg, Ca, Sr, Ba, and Ra) are called the **alkaline earth metals**. Elements of Group ⅥA (O, S, Se, Te, and Po) are sometimes referred to as the **chalcogens**. Elements in Group ⅦA (F, Cl, Br, I, and At) are known as the **halogens**, and elements in Group ⅧA (He, Ne, Ar, Kr, Xe, and Rn) are called the **noble gases**. The elements in Group ⅢB to ⅦB, Group Ⅷ, and Group ⅠB to ⅡB collectively are called the **transition elements** or the **transition metals**.

New Words and Expressions

actinide /ˈæktɪnaɪd/ n. 锕系元素
alkali metal n. 碱金属
alkaline earth metal n. 碱土金属
atomic number n. 原子序数
chalcogen /ˈkælkədʒən/ n. 氧族元素
group /ɡruːp/ n. 族
lanthanide /ˈlænθənaɪd/ n. 镧系元素

metalloid /ˈmetlɔɪd/ n. 类金属
noble gas n. 稀有气体
nonmetal /ˌnɒnˈmetəl/ n. 非金属
period /ˈpɪərɪəd/ n. 周期
periodic table n. 元素周期表
transition elements n. 过渡元素
transition metals n. 过渡金属

Notes

1. -oid 表示"像……的",如 metalloid（类金属）,steroid（类固醇）,humanoid（类人动物）。

2. non- 表示"非、不",如 nonmetal（非金属）,nonequilibrium（非平衡）,nonexistence（不存在）。

3. All atoms with the same number of protons are atoms of the same element and they have the same atomic number, Z.

参考译文：具有相同质子数的原子是同一种元素的原子，它们都有相同的原子序数 Z。

4. The elements are arranged in periods, horizontal rows, in order of increasing atomic number (the number of protons in the nucleus of an atom).

参考译文：元素按周期（横排）排列，并以原子序数（原子核中质子的数目）递增的顺序排列。

5. A metal is a good conductor of heat and electricity, whereas a nonmetal is usually a poor conductor of heat and electricity.

参考译文：金属是热和电的优良导体，而非金属通常是热和电的不良导体。

> **Questions**

1. What is the periodic table, and what is its significance in study of chemistry?
2. How does the periodic table be arranged?
3. Write the names and symbols for three elements in each of the following categories: (a) metal, (b) nonmetal, (c) metalloid.
4. Give two examples of each of the following: (a) transition metals, (b) alkali metals, (c) halogens, (d) alkaline earth metals, (e) chalcogens, (f) noble gases.

1.5 Chemical Reactions in Aqueous Solution

Most of the reactions in aqueous solution we study belong to one of these three types:

(1) **Precipitation reactions.** When two ionic solutions are mixed, a solid ionic substance in the form of precipitate is formed.

(2) **Acid-base reactions.** An acid reacts with a base. Such reactions involve the transfer of a proton from the Arrhenius acid to the Arrhenius base.

(3) **Oxidation-reduction reactions.** These involve the transfer of electrons between reactants.

1.5.1 Precipitation Reactions

A precipitation reaction occurs in aqueous solution because one product is insoluble. The insoluble solid compound is called a precipitate. To predict whether a precipitate will form when you mix two solutions of ionic compounds, you need to know whether any of the potential products that might form are insoluble. For example, we mix the solutions of calcium chloride, $CaCl_2$, and silver nitrate, $AgNO_3$. The balanced equation (assuming there is a reaction) can be written as

$$CaCl_2 + 2AgNO_3 \longrightarrow 2AgCl + Ca(NO_3)_2$$

In a precipitate reaction, the anions exchange between the two cations (or vice versa).

$CaCl_2$, $AgNO_3$, and $Ca(NO_3)_2$ are soluble in water but $AgCl$ is not. Thus, we predict that reaction occurs and the reaction can be written as

$$CaCl_2(aq) + 2AgNO_3(aq) \longrightarrow 2AgCl(s) + Ca(NO_3)_2(aq)$$

Here, (aq) and (s) stand for the states of aqueous solution and solid, respectively. This equation is called a **molecular equation.**

Next, we write the strong electrolytes (the highly soluble ionic compounds) in the form of ions, leaving the formula of the precipitate unchanged. This equation is called an **ionic equation.**

$$Ca^{2+}(aq) + 2Cl^-(aq) + 2Ag^+(aq) + 2NO_3^-(aq) \longrightarrow 2AgCl(s) + Ca^{2+}(aq) + 2NO_3^-(aq)$$

After canceling spectator ions (those that don't really react and that appear in an unchanged form on both sides of the equation) and reducing the coefficient to the smallest whole number, we obtain the **net ionic equation**

$$Ag^+(aq) + Cl^-(aq) \longrightarrow 2AgCl(s)$$

This equation represents the essential reaction that occurs between Ag^+ ions and Cl^- ions in aqueous solution. The product is the white precipitate, silver chloride.

1.5.2 Acid-Base Reactions

According to Arrhenius's definition, an acid is a substance that produces hydrogen ions, H^+, when it dissolves in water and a base is a substance that produces hydroxide ions, OH^-, when it dissolves in water. When a solution of an acid and that of a base are mixed, a **neutralization** reaction occurs. For example, when hydrochloric acid is mixed with sodium hydroxide, the following reaction occurs

$$HCl(aq) + NaOH(aq) \longrightarrow NaCl(aq) + H_2O(l)$$

Here, water and table salt, NaCl, are the products of the reaction. By analogy to this reaction, the term **salt** has come to mean any ionic compound whose cation comes from a base (Na^+ from NaOH) and the anion comes from an acid (Cl^- from HCl). In general, a neutralization reaction between an acid and a metal hydroxide produces water and a salt. The net ionic equation is

$$H^+(aq) + OH^-(aq) \longrightarrow H_2O(l)$$

This equation summarizes the essential features of the neutralization reaction between any strong acid and any strong base: H^+ and OH^- ions combine to form H_2O.

1.5.3 Oxidation-Reduction Reactions

Oxidation-reduction reactions (also called **redox reactions**) are reactions involving a transfer of electrons from one chemical species to another. Let's look at what happens when you dip a piece of iron plate into a blue solution of copper (Ⅱ) sulfate, $CuSO_4$. What you see is that the iron plate becomes coated with a reddish-brown tinge of metallic copper. The molecular equation for this reaction is

$$Fe(s) + CuSO_4(aq) \longrightarrow FeSO_4(aq) + Cu(s)$$

The net ionic equation is

$$Fe(s) + Cu^{2+}(aq) \longrightarrow Fe^{2+}(aq) + Cu(s)$$

Note that each iron atom in the metal loses two electrons to form iron (Ⅱ) ion, and each copper (Ⅱ) ion gains two electrons to form a copper atom on the metal. The net effect is that two electrons are transferred from each iron atom in the metal to each copper (Ⅱ) ion.

We can describe this reaction with two **half-reactions**. One of them involves a loss of electrons (or increase of oxidation number) and the other involves a gain of electrons (or decrease of oxidation number). The half-reactions are

$$Fe(s) \longrightarrow Fe^{2+}(aq) + 2e^- \quad (oxidation: electrons\ lost\ by\ Fe)$$
$$Cu^{2+}(aq) + 2e^- \longrightarrow Cu(s) \quad (reduction: electrons\ gained\ by\ Cu^{2+})$$

An **oxidizing agent** (oxidant) is a species that oxidizes another species, it is itself reduced. Similarly, a **reducing agent** (reductant) is species that reduces another species, it is itself oxidized. In this example reaction, the copper (Ⅱ) ion is the oxidizing agent, whereas iron metal is the reducing agent.

New Words and Expressions

acid-base reaction n. 酸碱反应
acid /ˈæsɪd/ n. 酸
anion /ˈænaɪən/ n. 阴离子
aqueous solution n. 水溶液
base /beɪs/ n. 碱
cation /ˈkætaɪən/ n. 阳离子
dissolve /dɪˈzɒlv/ v. 溶解
electrolyte /ɪˈlektrəlaɪt/ n. 电解质
half-reaction n. 半反应
hydrogen ion n. 氢离子
hydroxide ion n. 氢氧离子
ionic /aɪˈɒnɪk/ adj. 离子的
ionic equation n. 离子方程式
molecular equation n. 分子方程式

neutralization /ˌnjuːtrəlaɪˈzeɪʃn/ n. 中和
oxidant /ˈɒksɪdənt/ n. 氧化剂
oxidation-reduction reaction 氧化还原反应
oxidizing agent n. 氧化剂
precipitate /prɪˈsɪpɪteɪt/ n. 沉淀物
precipitation reaction n. 沉淀反应
reactant /riˈæktənt/ n. 反应物
reaction /riˈækʃn/ n. 化学反应
redox reaction n. 氧化还原反应
reducing agent n. 还原剂
reductant /rɪˈdʌktənt/ n. 还原剂
salt /sɔːlt/ n. 盐
soluble /ˈsɒljəbl/ adj. 可溶解的

Notes

1. We write the strong electrolytes (the highly soluble ionic compounds) in the form of ions, leaving the formula of the precipitate unchanged.

参考译文：我们以离子的形式表示强电解质（极易溶的离子化合物），而保持沉淀物的分子式不变。

2. After canceling spectator ions (those that don't really react and that appear in an unchanged form on both sides of the equation) and reducing the coefficient to the smallest whole number, we obtain the net ionic equation.

参考译文：在抵消旁观者离子（那些没有真正发生反应且在反应式两边都以相同形式出现的离子）后，并将系数降至最小的整数，可以得到净离子方程。

3. According to Arrhenius's definition, an acid is a substance that produces hydrogen ions, H^+, when it dissolves in water and a base is a substance that produces hydroxide ions, OH^-, when it dissolves in water.

参考译文：根据阿伦尼乌斯的定义，酸是一种溶解在水中后能产生氢离子（H^+）的物质，而碱是一种溶解在水中后能产生氢氧根离子（OH⁻）的物质。

4. Oxidation-reduction reactions are reactions involving a transfer of electrons from one chemical species to another.

参考译文：氧化还原反应是电子从一种化学物质转移到另一种化学物质的反应。

Questions

1. What is the difference between a molecular equation and an ionic equation? Between an ionic equation and a net ionic equation? What is the advantage of writing net ionic equations?

2. What are the products of an acid-base neutralization reaction?

3. Use this reaction, $4Na(s) + O_2(g) \longrightarrow 2Na_2O(s)$, to define the terms redox reaction, half-reaction, oxidizing agent, and reducing agent.

Chapter 2

Inorganic Chemistry

2.1 Nomenclature of Inorganic Compounds

In general, inorganic compounds are defined as compounds composed of elements other than carbon. Some exceptions, for example, include CO, CO_2, $CaCO_3$, and KCN, which contain carbon and yet are considered to be inorganic compounds. The following introduces the nomenclature of some simple ionic compounds and molecular compounds, respectively.

2.1.1 Naming Ionic Compounds

Most ionic compounds are substances composed of metals and nonmetals. The metals form the positive ions (called cations) and the nonmetals form the negative ions (called anions). We name an ionic compound by giving the name of the cation followed by the name of the anion. For instance, NaCl is named sodium chloride.

A monatomic ion is an ion formed from a single atom. Table 2.1 lists some common monatomic ions.

Table 2.1 Some common monatomic ions

Symbol	Name	Symbol	Name
H^+	Hydrogen ion	Pb^{2+}	Lead(Ⅱ) ion
Li^+	Lithium ion	Sn^{2+}	Tin(Ⅱ) ion
Na^+	Sodium ion	F^-	Fluoride ion
K^+	Potassium ion	Cl^-	Chloride ion
Ag^+	Silver ion	Br^-	Bromide ion
Mg^{2+}	Magnesium ion	I^-	Iodide ion
Ca^{2+}	Calcium ion	Sr^{2+}	Strontium ion
Zn^{2+}	Zinc ion	Cr^{2+}	Chromium(Ⅱ) ion

续表

Symbol	Name	Symbol	Name
Cr^{3+}	Chromium(III) ion	Hg^{2+}	Mercury(II) ion
Fe^{2+}	Iron(II) ion	Al^{3+}	Aluminum ion
Fe^{3+}	Iron(III) ion	H^-	Hydride ion
Co^{2+}	Cobalt(II) ion	O^{2-}	Oxide ion
Co^{3+}	Cobalt(III) ion	S^{2-}	Sulfide ion
Mn^{2+}	Manganese(II) ion	N^{3-}	Nitride ion

2.1.1.1 Rules for Naming Monatomic Ions

(1) Monatomic cations are named after the element if there is only one such ion. For example, K^+ is named potassium ion and Mg^{2+} is named magnesium ion.

(2) If there is more than one monatomic cation of an element, Roman numeral in parentheses denoting the charge on the ion. For example, Fe^{2+} is named iron(II) ion and Fe^{3+} is named iron(III) ion. There is an older system of nomenclature, in which ions are named by adding the suffixes *-ous* and *-ic* to a stem of the element to represent the ions of lower and higher oxidation state, respectively. For example, Fe^{2+} is named as ferrous ion and Fe^{3+} is named as ferric ion.

(3) The names of the monatomic anions are obtained from a stem of the element followed by the suffix *-ide*. For example, F^- is named as fluoride ion, from the stem name *fluor-* from fluorine and the suffix *-ide*.

A polyatomic ion is an ion consisting of two or more atoms chemically bonded together and carrying a net electric charge. Some common polyatomic ions are listed in Table 2.2. H_3O^+, NH_4^+, and Hg_2^{2+} are cations; the rest are anions.

Table 2.2 Some common polyatomic ions

Formula	Name	Formula	Name
H_3O^+	Hydronium ion	CH_3COO^-	Acetate ion
NH_4^+	Ammonium ion	CrO_4^{2-}	Chromate ion
Hg_2^{2+}	Mercury(I) ion	$Cr_2O_7^{2-}$	Dichromate ion
CN^-	Cyanide ion	NO_2^-	Nitrite ion
OH^-	Hydroxide ion	NO_3^-	Nitrate ion
$C_2O_4^{2-}$	Oxalate ion	PO_4^{3-}	Phosphate ion
MnO_4^-	Permanganate ion	HPO_4^{2-}	Hydrogen phosphate ion
CO_3^{2-}	Carbonate ion	$H_2PO_4^-$	Dihydrogen phosphate ion
HCO_3^-	Hydrogen carbonate ion(bicarbonate ion)	SO_3^{2-}	Sulfite ion
ClO^-	Hypochlorite ion	HSO_3^-	Hydrogen sulfite ion(bisulfite ion)
ClO_2^-	Chlorite ion	SO_4^{2-}	Sulfate ion
ClO_3^-	Chlorate ion	HSO_4^-	Hydrogen sulfate ion(bisulfate ion)
ClO_4^-	Perchlorate ion	$S_2O_3^{2-}$	Thiosulfate ion

From Table 2.2, we can see that

a. Very few polyatomic anions (CN^- and OH^-) have the *-ide* ending in their names.

b. The names of some oxoanions, which consist of oxygen with another element (called characteristic or central element), have a stem name from the characteristic element, plus a suffix *-ate* or *-ite*. The names of the oxoanions with the greater and lesser number of oxygen atoms have the suffix *-ate* and *-ite*, respectively. For example, SO_4^{2-} is named as sulfate ion; SO_3^{2-} is named as sulfite ion. Another example is NO_3^- (nitrate ion) and NO_2^- (nitrite ion).

c. If there are more than two oxoanions of a given characteristic element, some modifications are utilized. In the case of ClO^-, ClO_2^-, ClO_3^-, and ClO_4^-, prefixes *hypo-* and *per-* are used in addition to the two suffixes. The oxoanions having the least number of oxygen atoms, ClO_2^- and ClO^-, are named using the suffix *-ite*, and the prefix *hypo-* is added to the one of these two ions with the fewer oxygen atoms. Thus, ClO_2^- is named as chlorite ion, and ClO^- is named as hypochlorite ion. The oxoanions having the greatest number of oxygen atoms, ClO_3^- and ClO_4^-, are named using the suffix *-ate*, and the prefix *per-* is added to the one of these two ions with the greater oxygen atoms.

d. Some oxoanions listed in Table 2.2 are bonded to one or more hydrogen ions. They are named by adding as a prefix the word hydrogen or dihydrogen. For instance, HPO_4^{2-} is named as hydrogen phosphate ion, and $H_2PO_4^-$ is named as dihydrogen phosphate ion. It should be noticed that HCO_3^- (hydrogen carbonate ion) and HSO_3^- (hydrogen sulfite ion) are named as bicarbonate ion and bisulfite ion, respectively, in an older terminology.

2.1.1.2 *Naming Hydrates*

A hydrate is a compound that contains water molecules weakly bound in its crystals. Hydrates are named from the anhydrous compound, followed by the word **hydrate** with a prefix to indicate the number of water molecules. For example, $FeSO_4 \cdot 7H_2O$ is named as iron (Ⅱ) sulfate heptahydrate, and $CuSO_4 \cdot 5H_2O$ is named as copper (Ⅱ) sulfate pentahydrate.

2.1.2 Naming Binary Molecular Compounds

A binary compound is a compound composed of only two elements. Binary compounds consist of a metal and a nonmetal are named as ionic compounds as discussed above. Binary compounds compose of two nonmetals or metalloids are usually molecular and are named utilizing a prefix system.

2.1.2.1 *Rules for Naming Binary Molecular Compounds*

(1) The order of the compound usually has the elements in the order given in the formula.

(2) The first element in the formula is named using the exact element name.

(3) The second element in the formula is named by writing the stem name of the element with the suffix *-ide*.

(4) Prefixes, derived from the Greek, are used to denote the number of atoms of each element (the Greek prefixes are listed in Table 2.3). In general, the prefix *mono-* is not used, unless it is needed to distinguish two compounds of the same two elements.

Table 2.3 Greek prefix for naming compounds

Number	Prefix	Number	Prefix
1	Mono-	6	Hexa-
2	Di-	7	Hepta-
3	Tri-	8	Octa-
4	Tetra-	9	Nona-
5	Penta-	10	Deca-

Consider P_4S_{10}. According to Rule (1), we follow the order of the elements in the formula: P before S. By Rule (2), the P is named exactly as the element (phosphorus), and by Rule (3), the S is named as the anion (sulfide). The compound is phosphorus sulfide. Finally, following Rule (4), we add prefixes to denote the number of atoms of each element. As Table 2.3 shows, the prefix for number 4 is *tetra-*, and the prefix for number 10 is *deca-*. The name of the compound is tetraphosphorus decasulfide. Table 2.4 gives some other examples of binary molecular compounds. When the prefix ends in *a* or *o* and the element name begins with *a* or *o*, the final vowel of the prefix is dropped for ease of pronunciation. For example, monooxide becomes monoxide, and tetraoxide becomes tetroxide.

Table 2.4 Some examples of binary molecular compounds

Formula	Name	Formula	Name
BCl_3	Boron trichloride	NO_2	Nitrogen dioxide
CCl_4	Carbon tetrachloride	N_2O_4	Dinitrogen tetroxide
CO	Carbon monoxide	P_2O_5	Diphosphorus pentoxide
CO_2	Carbon dioxide	PCl_3	Phosphorus trichloride
NO	Nitrogen monoxide	SF_6	Sulfur hexafluoride

Here are some examples to illustrate how to use the prefix *mono-*. There is only one compound of hydrogen and bromine: HBr. It is called hydrogen bromide, not monohydrogen monobromide, since the prefix *mono-* is not generally written. On the other hand, there are two compounds of carbon and oxygen: CO and CO_2. They are both carbon oxides. To distinguish one from the other, the names of carbon monoxide and carbon dioxide are named for CO and CO_2, respectively.

Some compounds have older, well-established names. The compound H_2S would be named dihydrogen sulfide by the prefix system, but it is commonly called hydrogen sulfide. The names for H_2O and NH_3 are water and ammonia, not dihydrogen monoxide and trihydrogen mononitride, respectively.

2.1.2.2 *Naming Acids*

Even though we use names like hydrogen chloride for pure binary molecular compounds, we sometimes want to emphasize that their aqueous solution are acids. An acid is a substance that produces hydrogen ions when it dissolved in water. Binary acids are certain compounds of hydrogen with nonmetal atoms. They are named like binary compounds by using the prefix *hydro-* and the suffix *-ic* with the stem name of the nonmetal, followed by the word *acid*. Table 2.5 lists the most important binary acids.

Table 2.5 Some examples of binary acids

Acid	Name	Acid	Name
HF(aq)	Hydrofluoric acid	HI(aq)	Hydroiodic acid
HCl(aq)	Hydrochloric acid	H_2S(aq)	Hydrosulfuric acid
HBr(aq)	Hydrobromic acid		

The majority of acids are ternary compounds, which contain three different elements: hydrogen, and two other elements. If one of the nonmetals is oxygen, the acid is called an oxoacid. The oxoacids produce hydrogen ions and oxoanions in water. The names of the oxoacids are related to the names of the corresponding oxoanions. If the suffixes of the oxoanions are -*ate* and -*ite*, the suffixes of the oxoacids are -*ic* and -*ous*, respectively. Some examples of oxoacids are listed in Table 2.6.

Table 2.6 Some oxoanions and their corresponding oxoacids

Oxoanion	Name of Oxoanion	Oxoacid	Name of oxoacid
CO_3^{2-}	Carbonate ion	H_2CO_3	Carbonic acid
PO_4^{3-}	Phosphate ion	H_3PO_4	Phosphoric acid
NO_2^-	Nitrite ion	HNO_2	Nitrous acid
NO_3^-	Nitrate ion	HNO_3	Nitric acid
SO_3^{2-}	Sulfite ion	H_2SO_3	Sulfurous acid
SO_4^{2-}	Sulfate ion	H_2SO_4	Sulfuric acid
ClO^-	Hypochlorite ion	HClO	Hypochlorous acid
ClO_2^-	Chlorite ion	$HClO_2$	Chlorous acid
ClO_3^-	Chlorate ion	$HClO_3$	Chloric acid
ClO_4^-	Perchlorate ion	$HClO_4$	Perchloric acid

New Words and Expressions

binary /ˈbaɪnəri/ adj. 二元的
hydrate /haɪˈdreɪt/ n. 水合物
ionic compound n. 离子化合物
molecular compound n. 分子化合物
monatomic ion n. 单原子离子

nomenclature /nəˈmeŋklətʃə(r)/ n. 命名法
prefix /ˈpriːfɪks/ n. 前缀
suffix /ˈsʌfɪks/ n. 后缀

Notes

1. bi- 表示"二",如 binary(二元的),bidentate(二齿的),bicycle(脚踏车),biannual(一年两次的)。

2. 区分 HX 与 HX(aq)的英文命名(X 为卤素),例如 HF 为 hydrogen fluoride(氟化氢),而 HF(aq)为 hydrochloric acid(氢氟酸)。后者是在水溶液中,强调酸的特性。

> **Questions**

1. Name the following compounds or ions: (1) NF_3, (2) $SiCl_4$, (3) $Ca_3(PO_4)_2$, (4) BrO_4^-, (5) $MgSO_4 \cdot 7H_2O$, (6) Hg_2CrO_4, (7) $Fe_2(CO_3)_3$, (8) $Al(OH)_3$, (9) P_4O_6.
2. Name the following acids: (1) $HClO_3$, (2) HCl, (3) HF, (4) H_2SO_4.

2.2 Nomenclature of Coordination Compounds

Thousands of coordination compounds are now known. A systematic method of naming these compounds needs to provide the information about the structure of coordination compounds, such as the central metal, the chelating ligands, and the coordination number. The rules for naming coordination compounds, which are an extension of those originally given by Werner, are as follows:

(1) In names and formulas of coordination compounds, cation comes first, followed by anions. This is the same order as used in simple ionic compounds like sodium bromide, NaBr. For example, in complexes $K_4[Fe(CN)_6]$ and $[Co(NH_3)_6]Cl_3$, we named K^+ and $[Co(NH_3)_6]^{3+}$ first.

(2) Whether the complex ion bear a positive or negative charge, the name of the complex consists of two parts written together as one word. Ligands are named first, and the metal atom is named second.

$[Fe(CN)_6]^{4-}$ is named hexacyanoferrate (II) ion

$[Co(NH_3)_6]^{3+}$ is named hexaamminecobalt (III) ion

(3) The complete ligand name consists of a Greek prefix denoting the number of ligands, followed by the specific name of the ligand. When there are two or more ligands, the ligands were written in alphabetical order (disregarding Greek prefixes).

a. Anionic ligands end in *-o*. Normally, *ide* endings change to *o*, *ite* to *ito*, and *ate* to *ato* (see Table 2.7).

Table 2.7 Names of anionic and ligands

Anion name	Ligand name	Anion name	Ligand name
Fluoride, F^-	Fluoro	Sulfate, SO_4^{2-}	Sulfato
Chloride, Cl^-	Chloro	Thiosulfate, $S_2O_3^{2-}$	Thiosulfato
Bromide, Br^-	Bromo	Nitrite, NO_2^-	Nitrito-*N*-
Iodide, I^-	Iodo	Nitrite, ONO^-	Nitrito-*O*-
Oxide, O^{2-}	Oxo	Thiocyanate, SCN^-	Thiocyanato-*S*-
Hydroxide, OH^-	Hydroxo	Thiocyanate, NCS^-	Thiocyanato-*N*-
Cyanide, CN^-	Cyano		

b. Neutral ligands are usually given the name of molecule. There are several exceptions. Aqua, ammine, carbonyl, and nitrosyl are ligand names for H_2O, NH_3, CO, and NO, respectively.

c. The number of ligands of a given type is denoted by a prefix. The prefixes are *mono-* (1, usually omitted), *di-* (2), *tri-* (3), *tetra-* (4), *penta-* (5), *hexa-* (6), *hepta-* (7), *octa-* (8), *nona-* (9), and *deca-* (10).

d. When the name of the ligand also has a number prefix, the number of ligands is denoted with *bis-* (2), *tris-* (3), *tetrakis* (4), *pentakis-* (5), *hexakis-* (6), *heptakis-* (7), *octakis-* (8), *nonakis-* (9), and *decakis-* (10). The name of the ligand follows in parentheses. For example, $[Co(en)_6]Cl_3$ is named tris (ethylenediamine) cobalt (Ⅲ) chloride.

(4) The complete metal name consists of name of metal, followed by *-ate* if the complex is an anion, followed by the oxidation number of the metal as a Roman numeral in parentheses. (An oxidation state of zero is indicated by 0 in parentheses.) When there is a Latin name for the metal, it is usually used to name the anion. Table 2.8 lists the names for some metals in complex anions.

Table 2.8 Names for some metals in complex anions

Metal	Named in complex anions	Metal	Named in complex anions
Aluminum	Aluminate	Iron	Ferrate
Chromium	Chromate	Copper	Cuprate
Manganese	Manganate	Lead	Plumbate
Nickel	Nickelate	Silver	Argentate
Cobalt	Cobaltate	Tin	Stannate
Zinc	Zincate	Gold	Aurate
Molybdenum	Molybdate		

Some additional examples are listed below.

Complex	Name
$[CoCl(NH_3)_5]Cl_2$	Pentaamminechlorocobalt(Ⅱ) chloride
$[CrCl(NH_3)_5]SO_4$	Pentaamminechlorochrominum(Ⅲ) sulfate
$K_2[PtCl_6]$	Potassium hexachloroplatinate(Ⅳ)
$Na[AgF_4]$	Sodium tetrafluoroargentate(Ⅲ)
$K[Au(CN)_2]$	Potassium dicyanoaurate(Ⅰ)

New Words and Expressions

chelating ligand n. 螯合配体
coordination compound n. 配位化合物
coordination number n. 配位数
parenthesis /pəˈrenθəsɪs/ n. 圆括号

Questions

Naming the following coordination compounds: (1) $[Pt(NH_3)_4Cl_2]Cl_2$; (2) $K_3[FeF_6]$; (3) $(NH_4)_2[Fe(H_2O)F_5]$; (4) $[Fe(CO)_5]$; (5) $[Rh(CN)(en)_2]^+$; (6) $[Cr(NH_3)_6]_2[CuCl_4]_3$.

2.3 Ionic and Covalent Bonds

The properties of a substance, such as sodium chloride, are determined in part by the chemical bonds that held the atoms together. A **chemical bond** is a strong attractive force that exists between certain atoms in a substance. The reaction of sodium (a silvery metal) with chlorine (a pale greenish yellow gas) yields sodium chloride (a white solid). The substances in this reaction are quite different, as are their chemical bonds. Sodium chloride (NaCl) consists of sodium ions and chloride ions held in a regular arrangement, or crystal, by **ionic bonds**. Ionic bonding results from the attractive force between positively and negatively charged ions.

A second kind of chemical bond is a **covalent bond**. In a covalent bond, two atoms share valence electrons (outer-shell electrons), which are attracted to the positively charged cores of both atoms, thus linking them. For example, chlorine gas consists of Cl_2 molecules. It is not reasonable to expect the two Cl atoms in each Cl_2 molecule to acquire the opposite charges required for ionic bonding. Rather, a covalent bond holds the two atoms together. This is consistent with equal sharing of electrons that you would expect between identical atoms. In most molecules, the atoms are linked by covalent bonds.

There is the third kind of chemical bond, called **metallic bonding**, which exists in copper and other metals. A crystal of copper metal consists of a regular arrangement of copper atoms. The valence electrons of these atoms move throughout the crystal, attracted to the positive cores of all copper ions. This attraction holds the crystal together.

2.3.1 Ionic Bonds

The first explanation of chemical bonding was suggested by the special properties of salts, substances now known to be ionic. Salts are generally crystalline solids that melt at high temperatures. For example, sodium chloride melts at 801℃ and sodium bromide melts at 750℃. A molten salt, the liquid after melting, conducts an electronic current. A salt dissolved in water gives a solution that also conducts an electric current. The electrical conductivity of the molten salts and the salt solution results from the motion of ions in the liquids. This suggests the possibility that ions exist in certain solids, held together by the attraction between positively and negatively charged ions.

An ionic bond is a chemical bond formed by the electrostatic attraction between positive and negative ions. The bond forms between two atoms when one or more electrons are transferred from the valence shell of one atom to the valence shell of the other. The atom losing electrons becomes a cation (positive ion), and the atom gaining electrons becomes an anion (negative ion). Because of the electrostatic attraction any given ion tends to attract as many neighboring ions of opposite charge as possible. When large number of ions gather, they form an ionic solid. The solid normally has a regular, crystalline structure that allows for maximum attraction of ions, given their particular sizes.

In order to understand why ionic bonding occurs, consider the transfer of a valence electron from a sodium atom (electron configuration [Ne]$3s^1$) to the valence shell of a chlorine atom (electron configuration [Ne]$3s^23p^5$). As a result of the electron transfer between these two atoms, ions are formed, each of which has a noble-gas configuration. The sodium atom has lost its $3s$ electron and has taken on the helium configuration, [Ne]. The chlorine atom has accepted the electron into its $3p$ subshell and has taken on the argon configuration, [Ne]$3s^23p^6$. Such noble-gas configurations and the corresponding ions are particularly stable. This stability of these ions accounts in part for the formation of NaCl. Once a cation or an anion forms, it attracts ions of opposite charge. Within the sodium chloride crystal, NaCl, every sodium ion is surrounded by six chloride ions, and every chloride ion is surrounded by six sodium ions.

2.3.1.1 *Energy Involved in Ionic Bonding*

In a qualitative way, we have seen why a sodium atom and a chlorine atom might be expected to form an ionic bond. Let's consider the energy changes involved in ionic bond formation. From this analysis, we can gain further understanding of why certain atoms bond ionically and others do not.

If atoms come together and bond, there should be a net decrease in energy, because the bonded state should be more stable and therefore at a lower energy level. Consider again the information of an ionic bond between a sodium atom and a chlorine atom. You can think of this as occurring in two steps. First, an electron is transferred between the two separate atoms to give ions. Second, the ions then attract one another to form an ionic bond. In fact, the transfer of the electron and the formation of an ionic bond occur simultaneously, rather than in discrete steps, when the atom approach one another. But the net quantity of energy involved is the same whether the steps occur one after the other or at the same time.

The first step requires removal of the $3s$ electron from the sodium atom and the addition of this electron to the valence shell of the chlorine atom. Energy is required to remove the electron from the sodium atom. This is the first ionization energy of the sodium atom, which is equal to 496 kJ/mol. On the other hand, adding the electron to the chlorine atom releases energy. This is the negative of the electron affinity of the chlorine atom, which equals −349 kJ/mol. The overall energy of this step can be calculated to be (496−349) kJ/mol, or 147 kJ/mol (Figure 2.1, step 1). Thus, this process requires more energy to remove an electron from the sodium atom than is gained when the electron is added to the chlorine atom. The formation of ions from the atoms is not in itself energetically favorable.

When positive and negative ions bond, however, more than enough energy is released to make the overall process favorable. The attraction of oppositely charged ions determines the energy released when ions bond. Let us look first at the energy obtained when a sodium ion and a chloride ion come together to form an ion-pair molecule. We can estimate this energy by making the simplifying assumption that the ions are spheres, just touching with the distance between the nuclei of the ions equal to this distance in the NaCl crystal. From the experiments, this distance is measured to be 282 pm (or 2.82×10^{-10} m). Using Coulomb's law, we can calculate the energy obtained when the ion spheres come together.

Coulomb's law states that the potential energy obtained in bringing two charges Q_1 and Q_2, initially far apart, up to a distance r apart is directly proportional to the product of the charges and inversely proportional to the distance between them: $E = kQ_1Q_2/r$. Here k is a physical constant, equal to 8.99×10^9 J·m/C² (C is the symbol for coulomb). The charge on a sodium ion is $+e$ and that on a chloride ion is $-e$, where e equals 1.602×10^{-19} C. Thus, we can estimate that the energy of attraction of a sodium ion and a chloride ion to form an ion pair Na^+Cl^- is $E = -8.18 \times 10^{-19}$ J. The minus sign means energy is released. This energy is for the formation of one ion pair. To express this for one mole of Na^+Cl^- pairs, we multiple by Avogadro's number, 6.02×10^{23}. We obtain -493 kJ/mol for the energy obtained when one mole of sodium ions and one mole of chloride ions come together to form one mole of Na^+Cl^- ion pairs (Figure 2.1, step 2a).

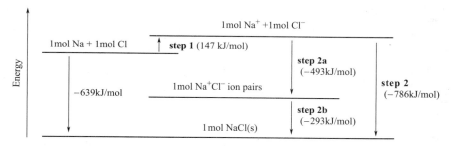

Figure 2.1 An enthalpy diagram for formation of NaCl

Now we see that the formation of ion pairs from sodium and chlorine atoms is energetically favorable. The attraction of oppositely charged ions, however, does not stop with the bonding of pairs of ions. The maximum attraction of ions of opposite charge with the minimum repulsion of ions of the same charge is obtained with the formation of the crystalline solid. Then additional energy is released. The energy of this step (Figure 2.1, step 2b) is most easily obtained as the difference between step 2a (Figure 2.1), which we just calculated, and step 2 (Figure 2.1), the energy released when a crystalline solid form the ions. This is the negative of the lattice energy of NaCl. Thus, the additional energy (in going from ion pairs to the crystalline solid) in step 2b (Figure 2.1) equals -293 kJ/mol.

The **lattice energy** is *the amount of energy required to convert a mole of ionic solid to its constituent ions in the gas phase*. For sodium chloride, the process is written as

$$NaCl(s) \longrightarrow Na^+(g) + Cl^-(g)$$

The distances between ions in the crystal are continuously enlarged until the ions are very far apart. You can obtain an experimental value for this process from thermodynamic data. The lattice energy for NaCl is 786 kJ/mol (Figure 2.1, step 2). Consequently, the net energy obtained when gaseous sodium and chlorine atoms form sodium chloride is $(-786+147) = -639$ kJ/mol. The negative sign shows that there has been a net decrease in energy, which you expect when stable bonding has occurred.

From this energy analysis, you can see that two elements bond ionically if the ionization energy of one is sufficiently small and the electron affinity of the other is sufficiently large. This situation exists between a reactive metal, which has low ionization energy, and a reactive nonmetal, which has large electron affinity. Generally, bonding between a metal and a nonmetal is ionic. This energy analysis also explains why ionic bonding normally results in a solid rather than in ion-pair molecules.

2.3.1.2 Lattice Energies from the Born-Haber Cycle

The preceding energy analysis requires the value of lattice energy of sodium chloride. Actually, direct experimental determination of the lattice energy of an ionic solid is difficult. However, this quantity can be indirectly determined from experiments by means of a thermochemical "cycle" originated by Max Born and Fritz Haber in 1919 and now called the Born-Haber cycle. The reasoning is based on Hess's law.

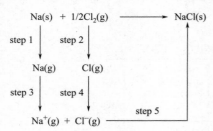

Figure 2.2　Born-Haber cycle for NaCl

To obtain the lattice energy of sodium chloride, you may think of solid sodium chloride being formed from the elements by two different routes, as shown in Figure 2.2. In route one, NaCl(s) is formed directly from the elements, Na(s) and $\frac{1}{2}Cl_2(g)$. The enthalpy change for this is $\Delta H_f^\ominus$, -411 kJ per mole of NaCl. The second route consists of the following five steps, along with the enthalpy changes for each (To be precise, the ionization energy and electron affinity are energy changes, ΔE, and we should add small correction to give the enthalpy changes, ΔH).

(1) Conversion of sodium metal into the gas phase sodium atoms. Metallic sodium is vaporized to a gas of sodium atoms (Sublimation is the transformation of a solid to a gas). The enthalpy change (ΔH) for this process, measured experimentally is 108kJ per mole of sodium.

(2) Cleavage of the Cl—Cl bond to convert Cl_2 molecules into chlorine atoms. Chlorine molecules are dissociated to chlorine atoms. The enthalpy change for this equals the Cl—Cl bond dissociation energy, which is 240 kJ per mole of bonds, or 120 kJ per mole of Cl atoms.

(3) Ionization of sodium to form sodium ions. Sodium atoms are ionized to Na^+ ions. The enthalpy change is essentially the ionization energy of atomic sodium, which equals 496 kJ per mole of Na.

(4) Addition of an electron to chlorine to form a chloride ion. The electrons from the ionization of sodium atoms are transferred to chloride atoms. The enthalpy change for this is the negative of the electron affinity of atomic chlorine, equal to -349 kJ per mole of Cl atoms.

(5) Combination of the gaseous ions to form the solid-state structure. The sodium ions and chloride ions formed in step 3 and 4 combine to give sodium chloride. Because this

process is just the reverse of the one corresponding to the lattice energy (breaking the solid into ions), the enthalpy change is the negative of the lattice energy. If we let U be the lattice energy, the enthalpy change for step 5 is $-U$ kJ/mol.

We can write these five steps and add them. In addition, we add the corresponding enthalpy change, following Hess's law.

$$Na(s) \longrightarrow Na(g) \qquad \Delta H_1 = 108 \text{ kJ/mol}$$

$$\frac{1}{2}Cl_2(g) \longrightarrow Cl(g) \qquad \Delta H_2 = 120 \text{ kJ/mol}$$

$$Na(g) \longrightarrow Na^+(g) + e^-(g) \qquad \Delta H_3 = 496 \text{ kJ/mol}$$

$$Cl(g) + e^- \longrightarrow Cl^-(g) \qquad \Delta H_4 = -349 \text{ kJ/mol}$$

$$Na^+(g) + Cl^-(g) \longrightarrow NaCl(s) \qquad \Delta H_5 = -U \text{ kJ/mol}$$

$$Na(s) + \frac{1}{2}Cl_2(g) \longrightarrow NaCl(s) \qquad \Delta H_f^\ominus = (375 - U) \text{ kJ/mol}$$

In summing the equations, we have canceled terms that appear on both the left and right sides of the arrows. The final equation is simply the formation reaction for NaCl(s) from sodium metal and chlorine gas. Adding the enthalpy changes, we find that the enthalpy change for this formation reaction is $(375-U)$ kJ/mol. The enthalpy of formation for NaCl(s) is -411 kJ/mol, determined calorimetrically. Equating these two values, we get

$$(375 - U) \text{ kJ/mol} = -411 \text{ kJ/mol}$$

Solving for U yields the lattice energy of NaCl

$$U = (375 + 411) \text{ kJ/mol} = 786 \text{ kJ/mol}$$

2.3.2 Covalent Bonds

The ionic substances are typically high-melting solids. However, many substances are molecular-gases, liquids, or low-melting solids consisting of molecules. A molecule is a group at atoms, frequently nonmetal atoms, strongly linked by chemical bonds. Often the forces that hold atoms together in a molecular substance cannot be understood on the basis of the attraction of oppositely charged ions (the ionic model). An obvious example is the molecule H_2, in which the two H atoms are held together tightly and no ions are present. In 1916, Gilbert Newton Lewis proposed that the strong attractive force between two atoms in a molecule results from a covalent bond, a chemical bond formed by the sharing of a pair of electrons between atoms. In 1926, Walter Heitler and Fritz London showed that the covalent bond in H_2 could be quantitatively explained by the newly discovered theory of quantum mechanics.

Let us consider the formation of a covalent bond between two H atoms to give the H_2 molecule. As the atoms approach one another, their $1s$ orbitals begin to overlap. Each electron can then occupy the space around both atoms. In other words, the two electrons can be shared by the H atoms. The electrons are attracted simultaneously by the positive charges of the two hydrogen nuclei. This attraction that bonds the electrons to both nuclei is the force holding the atoms together. Although ions do not exist in H_2, the force that holds the atom together can still be regarded as arising from the attraction of oppositely charged particles: nuclei and electrons.

It is interesting to see how the potential energy of the H atoms changes as they approach and then bond. As the H atoms approach, the potential energy gets lower and lower. The decrease in energy is a reflection of the bonding of atoms. Eventually, as the atoms get close enough, the repulsion of the positive charges on the nuclei becomes larger than the attraction of electrons for nuclei. In other words, the potential energy reaches a minimum value (-436 kJ/mol) and then increases. The distance (74 pm) between nuclei at this minimum energy is call the **bond length** of H_2. It is the normal distance between nuclei in the molecule.

Now imagine the reverse process. You start with the H_2 molecule, the H atoms at their normal bond length (74 pm) apart (at the minimum of the potential-energy curve). To separate the atoms in the molecule, energy must be added. The energy that must be added is called the **bond dissociation energy**. In the case of H_2 molecule, its bond dissociation energy is 436 kJ/mol. The larger the bond dissociation energy, the stronger the bond.

A covalent bond involves the sharing of at least one pair of electrons between two atoms. When the atoms are alike, as in the case of the H-H bond of H_2, the bonding electrons are shared equally. That is, the electrons spend the same amount of time in the vicinity of each atom. But when the two atoms are of different elements, the bonding electrons need not be shared equally. A **polar covalent bond** (or simply **polar bond**) is a covalent bond in which the bonding electrons spend more time near one atom than the other. For example, in the case of HBr molecule, the bonding electrons spend more time near the bromine atom than the hydrogen atom.

You can consider the polar covalent bond as intermediate between a nonpolar covalent bond, as in H_2, and an ionic bond, as in NaBr. From this viewpoint, an ionic bond is simply an extreme example of a polar covalent bond. The bonding pairs of electrons are equally shared in H_2, unequally shared in HBr, and essentially not shared in NaBr. Thus, it is possible to arrange different bonds to form a gradual transition from nonpolar covalent to polar covalent to ionic.

Note that a polar bond results when the bonding pair is drawn more toward one atom than the other. The concept of electronegativity is useful in judging whether a bond will be polar or not. Electronegativity is a measure of the ability of an atom in a molecule to draw bonding electrons to itself. Several electronegativity scales have been proposed. In 1934, Robert S. Mulliken suggested on theoretical grounds that the electronegativity (X) of an atom is given as half of its ionization energy ($I.E.$) plus electron affinity ($E.A.$). $X=(I.E.+E.A.)/2$.

An atom such as fluorine that tends to pick up electrons easily (large $E.A.$) and hold on to them strongly (large $I.E.$) has a large electronegativity. On the other hand, an atom such as lithium or cesium that loses electrons readily (small $I.E.$) and has little tendency to gain electrons (small $E.A.$) has a small electronegativity. Until recently, only a few electron affinities had been measured. For this reason, Mulliken's scale has had limited utility. A more widely used scale was derived earlier by Linus Pauling from bond enthalpies. Because electronegativities depend somewhat on the bonds formed, these values are actually average ones.

The absolute value of the difference in electronegativity of two bonded atoms gives a rough measure of the polarity to be expected in a bond. When this difference is small, the bond is nonpolar. When it is large, the bond is polar (or, if the difference is very large, perhaps ionic). The electronegativity differences for the bonds H—H, H—Br, and Na—Br are 0.0, 0.7, and 1.9, respectively, following the expected order. Differences in electronegativity can explain why ionic bonds usually form between a metal atom and a nonmetal atom; the electronegativity difference would be largest between these elements. On the other hand, covalent bonds form primarily between two nonmetals because the electronegativity differences are small.

You can use an electronegativity scale to predict the direction in which the electrons shift during bond formation; the electrons are pulled toward the more electronegative atom. For example, consider the H—Br bond. Because the Br atom ($X = 2.8$) is more electronegative than the H atom ($X = 2.1$), the bonding electrons in H—Br are pulled toward Br. Because the bonding electrons spend most of their time around the Br atom, that end of the bond requires a partial negative charge (indicated δ-). The H-atom end of the bond has a partial positive charge ($\delta+$). The HBr molecule is said to be a polar molecule.

New Words and Expressions

bond dissociation energy n. 键解离能
bond length n. 键长
Born-Haber cycle n. 波恩-哈伯循环
chemical bond n. 化学键
Coulomb's law n. 库仑定律
covalent bond n. 共价键
covalent /ˌkəʊˈveɪlənt/ adj. 共价的
electron affinity n. 电子亲和能
electron configuration n. 电子构型
electronegativity /ɪˌlektrəʊˌnegəˈtɪvɪti/ n. 电负性
Hess's law n. 赫斯定律

ionic /aɪˈɒnɪk/ adj. 离子的
ionic bond n. 离子键
ionization energy n. 电离能
lattice energy n. 晶格能
metallic bonding n. 金属键合
outer-shell electron n. 外层电子；价电子
polar covalent bond 极性共价键
simultaneously /ˌsɪmlˈteɪnɪəsli/ adv. 同时地
sublimation /ˌsʌblɪˈmeɪʃn/ n. 升华
valence electron n. 价电子
valence shell n. 价电子层；价层

Notes

1. The substances in this reaction are quite different, as are their chemical bonds.
参考译文：在这个反应中的物质和它们的化学键是完全不相同的。

2. We can estimate this energy by making the simplifying assumption that the ions are spheres, just touching with the distance between the nuclei of the ions equal to this distance

in the NaCl crystal.

参考译文：我们可以通过假设这些离子是互相接触的球体，并使用氯化钠晶体中正负离子的核间距来估算这个能量。

3. Coulomb's law states that the potential energy obtained in bringing two charges Q_1 and Q_2, initially far apart, up to a distance r apart is directly proportional to the product of the charges and inversely proportional to the distance between them: $E=kQ_1Q_2/r$.

参考译文：库仑定律指出，将两个电荷 Q_1 和 Q_2 从最初相距无限远移至距离为 r 时所获得的势能与电荷的乘积成正比，而与电荷之间的距离成反比：$E=kQ_1Q_2/r$。

4. An atom such as lithium or cesium that loses electrons readily (small $I.E.$) and has little tendency to gain electrons (small $E.A.$) has a small electronegativity.

参考译文：像锂或铯这样的原子很容易失去电子（即小的电离能），并且几乎没有获得电子的倾向（即小的电子亲和能），它的电负性很小。

Questions

1. Explain what ionic bonding is.
2. Explain what energy terms are involved in the formation of an ionic solid from atoms.
3. Describe the formation of a covalent bond in H_2 from H atoms.

2.4 Coordination Compounds

Ions of the transition elements exist in aqueous solution as **complex ions**. Iron(Ⅱ) ion, for example, exists in water as $[Fe(H_2O)_6]^{2+}$. The water molecules in this ion are arranged about the iron atom with their oxygen atoms bonded to the metal by donating electron pairs to it. Replacing the water molecules by six cyanate ions gives the $[Fe(CN)_6]^{4-}$ ion. Some of the transition elements have biological activity, and their role in human nutrition depends in most cases on the formation of **complexes**, or **coordination compounds**, which exhibit the type of bonding that occurs in $[Fe(H_2O)_6]^{2+}$ and $[Fe(CN)_6]^{4-}$.

2.4.1 Werner's Theory

Because compounds of the transition metals are beautifully colored, the chemistry of these elements fascinated chemists even before the periodic table was introduced. During the late 1700s through the 1800s, many coordination compounds that were isolated and studied had properties that were puzzling in light of the bonding theories prevailing at the time. Table 2.9 list a series of $CoCl_3$-NH_3 compounds that have strikingly different colors. Note that the third and fourth species have different colors even though the originally assigned formula was the same for both, $CoCl_3 \cdot 4NH_3$.

Table 2.9 Properties of some ammonia complexes of cobalt (Ⅲ)

Original formulation	Color	Ions per formula unit	"Free" Cl^- ions per formula unit	Modern formulation
$CoCl_3 \cdot 6NH_3$	orange	4	3	$[Co(NH_3)_6]Cl_3$
$CoCl_3 \cdot 5NH_3$	purple	3	2	$[Co(NH_3)_5Cl]Cl_2$
$CoCl_3 \cdot 4NH_3$	green	2	1	trans-$[Co(NH_3)_4Cl_2]Cl$
$CoCl_3 \cdot 4NH_3$	violet	2	1	cis-$[Co(NH_3)_4Cl_2]Cl$

The modern formulations of the compounds in Table 2.9 are based on various lines of experimental evidence. For example, all four compounds are strong electrolytes but yield different numbers of ions when dissolved in water. Dissolving $CoCl_3 \cdot 6NH_3$ in water yields four ions per formula unit ($[Co(NH_3)_6]^{3+}$ plus three Cl^- ions), whereas $CoCl_3 \cdot 5NH_3$ yields only three ions per formula unit ($[Co(NH_3)_5Cl]^{2+}$ and two Cl^- ions). Furthermore, the reaction of the compounds with excess aqueous silver nitrate leads to the precipitation of different amounts of $AgCl(s)$. When $CoCl_3 \cdot 6NH_3$ is treated with excess $AgNO_3(aq)$, 3 moles of $AgCl(s)$ are produced per mole of complex, which means all three Cl^- ions in the complex can react to form $AgCl(s)$. On the contrary, when $CoCl_3 \cdot 5NH_3$ is treated with excess $AgNO_3(aq)$, only 2 moles of $AgCl(s)$ precipitate per mole of complex are produced, telling us that one of the Cl^- ions in the complex does not react. These results are summarized in Table 2.9.

In 1893, the Swiss chemist Alfred Werner (1866—1919) proposed a theory that successfully explained the observations in Table 2.9. In a theory that became the basis for understanding coordination chemistry, Werner proposed that any metal ion exhibits both a primary valence and a secondary valence. The **primary valence** is the oxidation state of the metal, which is for the complexes in Table 2.9. The **secondary valence** is the number of atoms bonded to the metal ion, which is also called the **coordination number.** For these cobalt complexes, Werner deduced a coordination number of 6 with the ligands in an octahedral arrangement around the Co^{3+} ion.

Werner's theory provided a beautiful explanation for the results in Table 2.9. The amine molecules are ligands bonded to the ion; if there are fewer than six amine molecules, the remaining ligands are chloride ions. The central metal and the ligands bound to it constitute the coordination sphere of the complex.

In writing the chemical formula for a coordination compound, Werner suggested using square brackets to signify the makeup of the coordination sphere in any given compound. He therefore proposed that $CoCl_3 \cdot 6NH_3$ and $CoCl_3 \cdot 5NH_3$ are better written as $[Co(NH_3)_6]Cl_3$ and $[Co(NH_3)_5Cl]Cl_2$, respectively. He further proposed that the chloride ions that are part of the coordination sphere are bound so tightly that they do not dissociate when the complex is dissolved in water. Thus, dissolving $[Co(NH_3)_5Cl]Cl_2$ in water produces a $[Co(NH_3)_5Cl]^{2+}$ ion and two chloride ions.

Werner's ideas also explained why there are two forms of $CoCl_3 \cdot 4NH_3$. Using Werner's postulates, we write the formula as $[Co(NH_3)_4Cl_2]Cl$. As shown below, there are two ways to arrange the ligands in the $[Co(NH_3)_4Cl_2]^+$ complex, called the cis and *trans* forms. In the *cis* form, the two chloride ligands occupy adjacent vertices of the octahedral arrangement. In *trans*-[Co(NH$_3$)$_4$Cl$_2$]$^+$ the two chlorides are opposite each other. It is this difference in positions of the Cl ligands that leads to two compounds, one violet and one green.

$$\left[\begin{array}{c} \text{Cl} \diagdown \overset{\text{NH}_3}{\underset{\text{NH}_3}{|}} \diagup \text{NH}_3 \\ \text{Cl} \diagup \overset{\text{Co}}{\underset{|}{}} \diagdown \text{NH}_3 \end{array}\right]^+ \quad \left[\begin{array}{c} \text{H}_3\text{N} \diagdown \overset{\text{Cl}}{\underset{\text{Cl}}{|}} \diagup \text{NH}_3 \\ \text{H}_3\text{N} \diagup \overset{\text{Co}}{\underset{|}{}} \diagdown \text{NH}_3 \end{array}\right]^+$$

The insight Werner provided into the bonding in coordination compounds is even more remarkable when we realize that his theory predated Lewis's ideas of covalent bonding by more than 20 years! Because of his tremendous contributions to coordination chemistry, Werner was awarded the 1913 Nobel Prize in Chemistry.

2.4.2 Basic Definitions

A **complex ion** is a *metal ion with Lewis bases attached to it through coordinate covalent bonds*. A **complex** (or **coordination compound**) is *a compound consisting either of complex ions and other ions of opposite charge* (for example, the compound K$_3$[FeF$_6$] of the complex ion [FeF$_6$]$^{3-}$ and three K$^+$ ions) *or of a neutral complex species* (such as cisplatin, *cis*-[Pt(NH$_3$)$_2$Cl$_2$]).

Ligands (from the Latin word *ligare*, "to bind") are *the Lewis bases attached to the metal atom in a complex*. They are electron-pair donors, so ligands may be neutral molecules (such as H$_2$O or NH$_3$) or anions (such as CN$^-$ or Cl$^-$) that have at least one atom with a lone pair of electrons. Cations only rarely function as ligands. You might expect this, because an electron pair on a cation is held securely by the positive charge, so it would not be involved in coordinate bonding.

The ligands we have discussed so far bond to the metal atom through one atom of the ligand. For instance, ammonia bonds through the nitrogen atom. This type of bonding indicates a **monodentate ligand** (meaning "one-toothed" ligand), that is, *a ligand that bonds to a metal atom through one atom of the ligand*. A **bidentate ligand** ("two-toothed" ligand) is *a ligand that bonds to a metal atom through two atoms of the ligand*. Ethylenediamine (NH$_2$CH$_2$CH$_2$NH$_2$; it is frequently abbreviated in formulas as "en") is an example. The oxalate ion, C$_2$O$_4^{2-}$, is another common bidentate ligand.

Nitrogen atoms at the end of the molecule have lone pairs of electrons that can form coordinate covalent bonds. In forming a complex, for example, [Pt(en)$_2$]$^{2+}$, the ethylenediamine molecule bends around so that both nitrogen atoms coordinate to the metal atom, Pt.

$$\left[\begin{array}{c} \text{H}_2\text{C}-\text{H}_2\text{N} \diagdown \phantom{\text{Pt}} \diagup \text{NH}_2-\text{CH}_2 \\ \phantom{\text{H}_2\text{C}}|\phantom{\text{H}_2\text{N}} \text{Pt} \phantom{\text{NH}_2}|\phantom{\text{CH}_2} \\ \text{H}_2\text{C}-\text{H}_2\text{N} \diagup \phantom{\text{Pt}} \diagdown \text{NH}_2-\text{CH}_2 \end{array}\right]^{2+}$$

The hemoglobin molecule in red blood cells is an example of a complex with a **quadridentate ligand**-one that bonds to the metal atom through four ligand atoms. Hemoglobin consists of the protein globin chemically bonded to heme. Heme is a planar molecule consisting of iron (Ⅱ) to which a quadridentate ligand is bonded through its four nitrogen atoms.

Ethylenediaminetetraacetate ion (EDTA) is a **hexadentate ligand** that bonds through six of its atoms. It can completely envelop a metal atom, simultaneously occupying all six positions in an octahedral geometry.

$$\left[\begin{array}{c} O \\ \parallel \\ O-C-CH_2 \\ O-C-CH_2 \\ \parallel \\ O \end{array}\!\!N-CH_2-CH_2-N\!\!\begin{array}{c} O \\ \parallel \\ CH_2-C-O \\ CH_2-C-O \\ \parallel \\ O \end{array}\right]^{4-}$$

A **polydentate ligand** (having many teeth) is *a ligand that can bond with two or more atoms to a metal atom*. A complex formed by polydentate ligands is frequently quite stable and is called a **chelate**. Because of the stability of chelates, polydentate ligands (also called **chelating agents**) are often used to remove metal ions from a chemical system. EDTA, for example, is added to certain canned food. The same chelating agent has been used to treat lead poisoning because it binds Pb^{2+} ions as the chelate, forming a substance that can then be excreted by the kidneys.

The **coordination number** of a metal atom in a complex is *the total number of bonds the metal atom forms with ligands*. In $[Fe(H_2O)_6]^{2+}$, the iron atom bonds to each oxygen atom in the six water molecules. Therefore, the coordination number of iron in this ion is 6, by far the most common coordination number. Coordination number 4 is also well known, and many examples of number 5 have been discovered. The coordination number for an atom depends on several factors, but size of the metal atom is important. For example, coordination number 7 and 8 are seen primary in fifth- and sixth-period elements, whose atoms are relatively large.

New Words and Expressions

ammonia /ə'məuniə/ n. 氨
bidentate ligand n. 二齿配体
chelate /'kiːleɪt/ n. 螯合物
chelating agent n. 螯合剂
complex /'kɒmpleks/ n. 配合物
complex ion n. 配离子

coordination compound n. 配位化合物
coordination number n. 配位数
ligand /'lɪɡənd/ n. 配体
monodentate ligand n. 单齿配体
polydentate ligand n. 多齿配体
quadridentate ligand n. 四齿配体

Notes

1. poly- 表示"多",如 polydentate(多齿的), polymer(聚合物), polypeptide(多肽), polysaccharide(多糖), polyatomic(多原子的)。

2. mono- 表示"单个,一个",如 monodentate(单齿的), monomer(单体),

monologue（独白）。

3. quadri- 表示"四"，如 quadridentate（四齿的），quadricycle（四轮车），quadrilingual（用四种语言的）。

4. He further proposed that the chloride ions that are part of the coordination sphere are bound so tightly that they do not dissociate when the complex is dissolved in water.

参考译文：他进一步提出，作为配位层一部分的氯离子结合得非常牢固，以至于当配合物溶解在水中时，它们也不会离解。

5. A complex (or coordination compound) is a compound consisting either of complex ions and other ions of opposite charge.

参考译文：配合物（或配位化合物）是由配离子和其他带相反电荷的离子所组成的化合物。

Questions

1. Define the terms: complex ions, ligand, and coordination number. Use an example to illustrate the use of these terms.

2. Define the term: bidentate ligand. Give two examples.

2.5　The Halogens

The word halogen was introduced in 1811 to describe the ability of chlorine to form ionic compounds with metals. The name is based on the Greek words *halos* and *gen* meaning salt former. The name was later extended to include fluorine, bromine, and iodine as well. Mendeleev placed the halogens in Group Ⅶ of his periodic table, which is now group 17 in the IUPAC table. Current interest in the halogens extends far beyond their ability to form metallic salts.

2.5.1　Properties

The halogens exist as diatomic molecules, symbolized by X_2, where X is a generic symbol for a halogen atom. That these elements occur as nonpolar diatomic molecules accounts for their relatively low melting and boiling points. As expected, melting and boiling points increase as we move down the group from the smallest and lightest member of the group, fluorine, to the largest and heaviest, iodine. Conversely, chemical reactivity toward other elements and compounds increases in the opposite order, with fluorine being the most reactive and iodine the least reactive.

All the halogen atoms have large electron affinities and show a strong tendency to gain electrons. Consequently, the halogens are rather good oxidizing agents. The elements of the second period 2 have distinctly different chemistry from the rest of the group because of their small size and inability to expand their valence shells. However, for the halogens, the differences between the second-row element (fluorine) and the members of the group are much less dramatic. Still, fluorine differs from the other halogens in a few ways. For example, a fluorine atom almost always forms just one covalent bond, whereas chlorine, bromine, and iodine atoms typically form more than one bond and as many as seven in some of their compounds. Although all the halogens are quite reactive and are found in nature only as compounds, fluorine is considerably more reactive than the other members of the group. It reacts directly with almost all the elements, except for oxygen, nitrogen, and the lighter noble gases, and forms compounds with even the most unreactive metals. It reacts with almost all materials, especially organic compounds, to produce fluorides. The reactivity of fluorine can be attributed to the weakness of the fluorine-fluorine bond in F_2, which arises, as mentioned earlier, because of the small size of the fluorine atom and the repulsion between the lone pairs on the fluorine atoms.

Fluorine differs from the other halogens in that it shows a much greater tendency to form ionic bonds with metals. Perhaps this is most evident when we look at the binary compounds formed by the halogens and the group 13 metals. The trifluoride of Al, Ga, and In are all ionic compounds, with very high lattice energies and very high melting points ($>1000\ ℃$) whereas the trichlorides are volatile compounds with much lower melting points ($<600\ ℃$). For $AlCl_3$, $GaCl_3$, and $InCl_3$, the bonding is largely covalent because chloride ions are much larger, and much more polarizable, than fluoride ions. In addition, in the solid state, the chloride of the group 13 metals contains dimers, M_2Cl_6, whereas the fluorides of the group 13 metals are all ionic lattices containing M^{3+} and F^- ions.

Another important difference between fluorine and the other halogens is that fluorine shows the ability to stabilize other elements in very high oxidation states. For example, fluorine reacts with sulfur to give SF_6, with sulfur in the $+6$ oxidation state, whereas chlorine reacts directly with molten sulfur to give S_2Cl_2, with sulfur in the $+1$ oxidation state.

Much of the reaction chemistry of the halogens involves oxidation-reduction reactions in aqueous solutions. For these reactions, standard electrode potentials are helpful for understanding the reactivity of the halogens. Among the properties of the halogens are potentials for the following half-reaction:

$$X_2 + 2e^- \longrightarrow 2X^- (aq)$$

By this measure, fluorine is clearly the most reactive element of the group ($E^\ominus = 2.866\ V$). Of all the elements, it shows the greatest tendency to gain electrons and is therefore the most easily reduced. Thus, it is not surprising that fluorine occurs naturally only in combination with other elements, and only as the fluoride ion, F^-. Although both chlorine and bromine can exist in a variety of positive oxidation states, they are found in their naturally occurring compounds only as chloride and bromide ions. There are, however, naturally occurring compounds in which iodine is in a positive oxidation state (such as the iodate ion, IO_3^-, in $NaIO_3$). In the case of iodine, the tendency for I_2 to be reduced to I^- is not particularly great ($E^\ominus = 0.535\ V$).

When we summarize the reduction tendencies of main-group metals and their ions, generally one or, at most, a few $E^\ominus$ values tell the story, and these values are easily incorporated into tables. However, the oxidation-reduction chemistry of some of the nonmetals is much richer and involves a larger number of relevant $E^\ominus$ values. In these cases, electrode potential diagrams are particularly useful for summarizing $E^\ominus$ data. In these diagrams, a number written above a line segment is the $E^\ominus$ value for reduction of the species on the left (higher oxidation state) to the one on the right (lower oxidation state). For a reduction involving species not joined by a line segment, we generally can calculate the appropriate value of $E^\ominus$.

2.5.2 Preparation

Although the existence of fluorine had been known since early in the nineteenth century, no one was able to devise a chemical reaction to extract the free element from its compounds. Finally, in 1886, Moissan succeeded in preparing $F_2(g)$ by an electrolysis reaction. Moissan's method, which is still the only important commercial method for fluorine extraction, involves the electrolysis of HF dissolved in molten KHF_2. The chemical equation for the reaction is given below

$$2HF \xrightarrow{\text{elecrolysis}} H_2(g) + F_2(g)$$

The corresponding half-reactions are as follows

Anode: $\qquad 2F^-(aq) \longrightarrow F_2(g) + 2e^-$

Cathode: $\qquad 2H^+(aq) + 2e^- \longrightarrow H_2(g)$

Moissan also developed the electric furnace and was honored for both these achievements with the Nobel Prize in Chemistry in 1906. Nevertheless, the challenge of producing fluorine by means of a chemical reaction remained. In 1986, one century after Moissan isolated fluorine, the chemical synthesis of fluorine was announced.

Although chlorine can be prepared by several chemical reactions, electrolysis of NaCl (aq) is the usual industrial method. The electrolysis reaction is

$$2Cl^-(aq) + 2H_2O(l) \xrightarrow{\text{electrolysis}} 2OH^-(aq) + H_2(g) + Cl_2(g)$$

Bromine can be extracted from seawater, where it occurs in concentrations of about 70 ppm (1 ppm = 10^{-6}) as Br^-, or from inland brine sources. Seawater from the Dead Sea is a good source of bromine. The seawater or brine solution is adjusted to pH 3.5 and treated with $Cl_2(g)$, which oxidizes Br^- to Br_2 in the following displacement reaction

$$Cl_2(g) + 2Br^-(aq) \longrightarrow Br_2(l) + 2Cl^-(aq) \qquad E^\ominus_{cell} = 0.293 \text{ V}$$

The liberated Br_2 is swept from seawater with a current of air or from brine with steam. A dilute bromine vapor forms and can be concentrated by various methods. The reaction above also forms the basis of a test for the presence of Br^-.

Certain marine plants, such as seaweed, absorb and concentrate I^- selectively in the presence of Cl^- and Br^-. Iodine is obtained in small quantities from such plants. I_2 can be obtained from inland brines by a process similar to that for the production of Br_2. Another abundant natural source of iodine is NaIO$_3$, found in large deposits in Chile. Because the oxidation state of iodine must be reduced from $+5$ in IO$_3^-$ to 0 in I$_2$, the conversion of IO$_3^-$ to I$_2$ requires the use of a reducing agent. Aqueous sodium hydrogen sulfite (sodium bisulfite) is used as the reducing agent in the first part of a two-step procedure, followed by the reaction of I$^-$ with additional IO$_3^-$ to produce I$_2$. The net ionic equations for the reactions are given below.

$$IO_3^-(aq) + 3HSO_3^-(aq) \longrightarrow I^-(aq) + 3SO_4^{2-}(aq) + 3H^+(aq)$$
$$5I^-(aq) + IO_3^-(aq) + 6H^+(aq) \longrightarrow 3I_2(s) + 3H_2O(l)$$

2.5.3 Uses

The halogen elements form a variety of useful compounds, and the elements themselves are largely used to produce these compounds. All the halogens are used to make halogenated organic compounds. For example, elemental fluorine is used to produce compounds such as polytetrafluoroethylene, a plastic more commonly known as Teflon. In the past, fluorine was used to make chlorofluorocarbons (CFCs), which were used as refrigerants, but international treaties have banned the production of CFCs in most countries because they damage the stratospheric ozone layer. Now fluorine is used to make hydrochlorofluorocarbons (HCFCs), which are more environmentally benign alternatives to CFCs. Fluorinated organic compounds tend to be chemically inert, and it is this inertness that makes them useful as components in harsh chemical environments. Fluorine is a key element in a variety of useful inorganic compounds.

With an annual production of more than 13 million metric tons, elemental chlorine ranks about eighth in quantity among manufactured chemicals in the world. It has three main commercial uses: (1) production of chlorinated organic compounds (about 70%), chiefly ethylene dichloride, CH_2ClCH_2Cl, and vinyl chloride, $CH_2=CHCl$ (the monomer of polyvinyl chloride, PVC); (2) as a bleach in the paper and textile industries and for the treatment of swimming pools, municipal water, and sewage (about 20%); (3) production of dozens of chlorine-containing inorganic chemicals (about 10%).

Bromine is used to make brominated organic compounds. Some of these are used as fire retardants and pesticides. Others are used extensively as dyes and pharmaceuticals. An important inorganic bromine compound is AgBr, the primary light-sensitive agent used in photographic film.

Iodine is of much less commercial importance than chlorine. Iodine and its compounds, however, do have applications as catalysts, antiseptics, and germicides and in the preparation of pharmaceuticals and photographic emulsions (as AgI).

2.5.4 Hydrogen Halides

In aqueous solution, the hydrogen halides are called the hydrohalic acids. Except for HF, hydrohalic acids are strong acids in water.

One well-known property of HF is its ability to etch (and ultimately to dissolve) glass. The reaction is similar to one between HF and silica, SiO_2.

$$SiO_2(s) + 4HF(aq) \longrightarrow 2H_2O(l) + SiF_4(g)$$

Because HF reacts with glass, it must be stored in special containers coated with a lining of Teflon or polyethylene.

When a halide salt (such as fluorite, CaF_2) reacts with a nonvolatile acid [such as concentrated H_2SO_4 (aq)] in the presence of heat, a sulfate salt and the volatile hydrogen halide are produced.

$$CaF_2(s) + H_2SO_4(aq) \xrightarrow{\Delta} CaSO_4(s) + 2HF(g)$$

This method also works for preparing $HCl(g)$ but not $HBr(g)$ or $HI(g)$. Concentrated $H_2SO_4(aq)$ is a sufficiently strong oxidizing agent to oxidize Br^- to Br_2 and I^- to I_2. For example, the reaction of $NaBr(s)$ and concentrated $H_2SO_4(aq)$ yields $Br_2(g)$ and not $HBr(g)$.

$$2NaBr(s) + 2H_2SO_4(aq) \xrightarrow{\Delta} Na_2SO_4(s) + 2H_2O(l) + Br_2(g) + SO_2(g)$$

We can get around this difficult by using a nonoxidizing nonvolatile acid, such as phosphoric acid, H_3PO_4. In addition, all the hydrogen halides can be formed by direct combination of the elements, as shown below:

$$H_2(g) + X_2(g) \longrightarrow 2HX(g)$$

The reaction of $H_2(g)$ and $F_2(g)$ is very fast, however, occurring with explosive violence under some conditions. With $H_2(g)$ and $Cl_2(g)$, the reaction also proceeds rapidly (explosively) in the presence of light (photochemically initiated), although some HCl is made this way commercially. With Br_2 and I_2, the reaction occurs more slowly and a catalyst is required.

2.5.5 Oxoacids and Oxoanions of Chlorine

Fluorine, the most electronegative element, adopts the -1 oxidation state in its compounds. The other halogens when bonded to a more electronegative element such as oxygen, can have any one of several positive oxidation states: $+1$, $+3$, $+5$, $+7$. Chlorine forms a complete set of oxoacids in all these oxidation states, but bromine and iodine do not. Only a few of the oxoacids can be isolated in pure form ($HClO_4$, HIO_3, HIO_4, H_5IO_6); the rest are stable only in aqueous solution.

An easily prepared oxidizing agent for laboratory use is an aqueous solution of chlorine, which is called "chlorine water". The solution is not just one of Cl_2, however, because Cl_2(aq) disproportionates into $HOCl(aq)$ and $HCl(aq)$. Although the disproportionation of Cl_2 in water is nonspontaneous when all reactants and products are in their standard states, the reaction does not occur to a limited extent in solutions that are not strongly acidic.

Reduction: $Cl_2(g) + 2e^- \longrightarrow 2Cl^-(aq)$

Oxidation: $Cl_2(g) + 2H_2O(l) \longrightarrow 2HOCl(aq) + 2H^+(aq) + 2e^-$

Overall: $Cl_2(g) + H_2O(l) \longrightarrow Cl^-(aq) + HOCl(aq) + H^+(aq)$ $\quad E_{cell}^{\ominus} = -0.253\ V$

In contrast, the disproportionation is spontaneous for standard-state conditions in basic solution.

Reduction: $Cl_2(g) + 2e^- \longrightarrow 2Cl^-(aq)$

Oxidation: $Cl_2(g) + 4OH^-(aq) \longrightarrow 2OCl^-(aq) + 2H_2O(l) + 2e^-$

Overall: $Cl_2(g) + 2OH^-(aq) \longrightarrow Cl^-(aq) + OCl^-(aq) + H_2O(l)$ $\quad E_{cell}^{\ominus} = 0.937\ V$

An aqueous solution of HOCl is used as an effective germicide in the treatment of swimming pools and water purification. Aqueous solutions of hypochlorite salts, notably NaOCl(aq), are used as common household bleaches. Using an aqueous solution $Ca(OH)_2$ and a gaseous Cl_2 can make a solid household bleach, $Ca(OCl)Cl$, which contains both OCl^- and Cl^- ions.

ClO_2 (chlorine dioxide) is an important bleach for fibers and paper. Its reduction with peroxide ion in aqueous solution produces chlorite ions.

$$2ClO_2(g) + O_2^{2-}(aq) \longrightarrow 2ClO_2^-(aq) + O_2(g)$$

Sodium chlorite is used as a bleaching agent for textiles.

Chlorate salts containing the ClO_3^- ion form when $Cl_2(g)$ disproportionates in hot alkaline solutions.

$$3Cl_2(g) + 6OH^-(aq) \longrightarrow 5Cl^-(aq) + ClO_3^-(aq) + 3H_2O(l)$$

Chlorate are good oxidizing agents. In addition, solid chlorates produce oxygen gas when they decompose, which makes them useful in matches and fireworks. A simple laboratory method of producing $O_2(g)$ involves heating $KClO_3(s)$ in the presence of a catalyst, MnO_2.

$$KClO_3(s) \xrightarrow{\triangle} 2KCl(s) + 3O_2(g)$$

A similar reaction is used as a source of emergency oxygen in submarines and aircraft.

Perchlorate salts are prepared mainly by electrolysis of chlorate solutions. Oxidation of ClO_3^- occurs at a Pt anode through this half-reaction:

$$ClO_3^-(aq) + H_2O(l) \longrightarrow ClO_4^-(aq) + 2H^+(aq) + 2e^- \quad E^{\ominus} = -1.189\ V$$

Perchlorates are relatively stable compared with the other oxoacid salts. Because no oxidation state higher than +7 is available to chlorine, they do not disproportionate. At elevated temperatures or in the presence of a readily oxidizable compound, however, perchlorate salts may react explosively, so caution is advised when using them. Mixtures of powdered aluminum and ammonium perchlorate are used as the propellant in some solid-fuel rockets, such as those used on the space shuttle. Ammonium perchlorate is especially dangerous to handle because an explosive reaction may occur when the oxidant ClO_4^- acts on the reductant NH_4^+.

New Words and Expressions

antiseptic /ˌæntiˈseptɪk/ n. 防腐剂
brine /braɪn/ n. 盐水
emulsion /ɪˈmʌlʃn/ n. 乳浊液
etch /etʃ/ v. 蚀刻
germicide /ˈdʒɜːmɪsaɪd/ n. 杀菌剂

lining /ˈlaɪnɪŋ/ n. 衬层
metric /ˈmetrɪk/ adj. 米制的；公制的
pesticide /ˈpestɪsaɪd/ n. 杀虫剂
retardant /rɪˈtɑːd(ə)nt/ n. 阻燃剂
stratospheric ozone layer n. 平流层臭氧层

Notes

1. -cide 表示"切割",如 germicide(杀菌剂),insecticide(杀虫剂),decide(决定)。

2. As expected, melting and boiling points increase as we move down the group from the smallest and lightest member of the group, fluorine, to the largest and heaviest, iodine.

参考译文:与预期一致,(同一主族物质的)熔点和沸点随着同族向下移动而增加,从最小(原子序数最小)和最轻(原子量最低)的氟元素向下移动到最大(原子序数最大)和最重(原子量最高)的碘元素。

3. It reacts directly with almost all the elements, except for oxygen, nitrogen, and the lighter noble gases, and forms compounds with even the most unreactive metals.

参考译文:它与氧、氮和较轻的稀有气体不反应,但与其他元素(甚至最不活泼的金属)进行反应形成化合物。

4. The reactivity of fluorine can be attributed to the weakness of the fluorine-fluorine bond in F_2, which arises, as mentioned earlier, because of the small size of the fluorine atom and the repulsion between the lone pairs on the fluorine atoms.

参考译文:F_2 的反应性可归因于其弱的氟-氟键。如前所述,由于氟原子的小体积和氟原子的孤电子对之间的排斥力使得氟-氟键变弱。

5. Aqueous sodium hydrogen sulfite (sodium bisulfite) is used as the reducing agent in the first part of a two-step procedure, followed by the reaction of I^- with additional IO_3^- to produce I_2.

参考译文:亚硫酸氢钠在水中作为还原剂(两步反应的第一个反应),其后碘离子与另外的碘酸根反应生成碘(第二个反应)。

6. Fluorinated organic compounds tend to be chemically inert, and it is this inertness that makes them useful as components in harsh chemical environments.

参考译文:有机氟化合物往往具有化学惰性,因此它们可以被用在严苛的化学环境中。

Questions

1. Write a chemical equation to represent the reaction of (a) $Cl_2(g)$ with cold NaOH (aq), (b) $Cl_2(g)$ with Br^- (aq).

2. If Br^- and I^- occur together in aqueous solution, I^- can be oxidized to IO_3^- with an excess of $Cl_2(g)$. Simultaneously, Br^- is oxidized to Br_2, which is extracted with CS_2(l). Write chemical equations for reactions that occur.

Chapter 3

Organic Chemistry

3.1 What is Organic Chemistry?

Organic chemistry is the branch of chemistry involving organic compounds. In 1807, Jöns Jacob Berzelius defined an **organic compound** as one that could be obtained from a living organism, whereas **inorganic compounds** encompassed everything else. It was believed that inorganic compounds could be made in the laboratory, but organic compounds could not and only living systems could summon up a mysterious "vital force" needed to synthesize them. This belief was called **vitalism**. By this definition, any familiar compounds, such as glucose (a sugar), testosterone (a hormone), and deoxyribonucleic acid (DNA), are organic.

glucose   testosterone   royal purple

This definition of organic compounds broke down in 1828, when Friedrich Wöhler (1800—1882), a German physician and chemist, synthesized urea (an organic compound known to be a major component of mammalian urine) by heating an aqueous solution of ammonium cyanate (an inorganic salt).

$$(NH_4)^+(NCO)^- \xrightarrow{\text{heat}} H_2N-\underset{\underset{O}{\|}}{C}-NH_2$$

　　　　ammonium cyanate　　　　　　　urea

　　If vitalism couldn't account for the distinction between organic and inorganic compounds, what could? Gradually, chemists arrived at our modern definition: "An organic compound contains a substantial amount of carbon and hydrogen". This definition, however, is still imperfect because it leaves considerable room for interpretation. For example, many chemists could classify carbon dioxide (CO_2) as an inorganic compound because it does not contain any hydrogen atoms, whereas others would argue that it is an organic compound because it contains carbon and is critical in living systems. In plants, it is a starting material in photosynthesis, and in animals, it is a by-product of respiration. Similarly, tetrachloromethane (carbon tetrachloride, CCl_4) contains no hydrogen, but many would classify it as an organic compound. Butyllithium (C_4H_9Li), on the other hand, is considered by many to be an inorganic compound, despite the fact that 13 of 14 atoms are carbon or hydrogen. Although this definition of an organic compound has its inadequacies, it does allow chemists to classify most molecules.

　　The birth of organic chemistry as a distinct field occurred around the time that vitalism was dismissed, making the discipline less than 200 years old. However, humans have taken advantage of organic reactions and the properties of organic compounds for thousands of years! Since about 6000 BC, for example, civilizations have fermented grapes to make wine. Some evidence suggests that Babylonians, as early as 2800 BC, could convert oils into soaps.

　　Many clothing dyes are organic compounds. Among the most notable of these dyes is royal purple, also called Tyrian purple, which was obtained by ancient Phoenicians from a type of aquatic snail called Bolinus brandaris. However, these organisms produce this compound in small amounts: one gram of royal purple needs 10,000 of Bolinus brandaris to be processed.

　　Organic chemistry has matured tremendously since its inception. Today, we can not only use organic reactions to reproduce complex molecules found in natural, but also engineer new molecules never before seen.

New Words and Expressions

ammonium cyanate n. 氰酸铵
aquatic snail n. 水生螺（Bolinus brandaris 为其中的一种）
aqueous /ˈeɪkwɪəs/ adj. 水溶液的
Babylonian n. 巴比伦人
civilization /ˌsɪvəlaɪˈzeɪʃn/ n. 文明
considerable /kənˈsɪdərəbl/ adj. 相当多的

deoxyribonucleic acid (DNA) n. 脱氧核糖核酸
discipline /ˈdɪsəplɪn/ n. 学科
dismiss /dɪsˈmɪs/ v. 摒弃
distinction /dɪˈstɪŋkʃn/ n. 差别；区别
ferment /fəˈment/ v. （使）发酵
glucose /ˈɡluːkəʊs/ n. 葡萄糖

inadequacies /ɪnˈædɪkwəsi/ n. 不充分；不足（复数）
inception /ɪnˈsepʃn/ n. 开端；创始
interpretation /ɪnˌtɜːprəˈteɪʃn/ n. 解释
mammalian /məˈmeɪliən/ adj. 哺乳动物的
Phoenician /fəˈniːʃn/ n. 腓尼基人
photosynthesis /ˌfəʊtəʊˈsɪnθəsɪs/ n. 光合作用

respiration /ˌrespəˈreɪʃn/ n. 呼吸
royal purple n. 皇家紫
testosterone /teˈstɒstərəʊn/ n. 睾酮
Tyrian purple n. 泰尔紫
urea /jʊˈriːə/ n. 尿素
urine /ˈjʊərɪn/ n. 尿
vital force n. 生命力
vitalism /ˈvaɪtəˌlɪzəm/ n. 活力（生机）论

Questions

What is the definition of "organic compounds"?

3.2 Classes of Organic Compounds

There are more than 24 million organic molecules known in the world. Different types of organic compounds, which have their own characteristic properties and reactivities, result from the following:

(1) Carbon's ability to form chains by bonding with itself.
(2) The presence of element other than carbon and hydrogen.
(3) Functional groups.
(4) Multiple bonds.

Let us consider two isomers with the same molecular formula C_3H_6O:

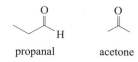

propanal　　acetone

Although the two isomers contain exactly the same type and number of atoms, their different arrangements of atoms result in two very different compounds. The first is an **aldehyde** called propanal; the second is a **ketone** called acetone. Aldehydes and ketones are two classes of organic compounds.

A class of organic compounds often is represented with a general formula that shows the atoms of the functional group(s) explicitly, and the remainder of the molecule using one or more Rs, where R represents an alkyl group. An **alkyl group**, a portion of an alkane, is formed by removing one hydrogen atom from the corresponding alkane. For instance, the methyl group (—CH_3), is formed by removing a hydrogen atom from the simplest alkane, methane (CH_4). Methyl groups are found in many organic molecules. Table 3.1 lists some of the simplest alkyl groups. Table 3.2 gives the general formula for each of the classes of selected organic compounds.

Table 3.1 Some alkyl groups

Name	Formula
Methyl（甲基）	—CH_3
Ethyl（乙基）	—CH_2CH_3
Propyl（丙基）	—$CH_2CH_2CH_3$
Isopropyl（异丙基）	—$CH(CH_3)_2$
Butyl（丁基）	—$CH_2CH_2CH_2CH_3$
tert-butyl（叔丁基）	—$C(CH_3)_3$
Pentyl（戊基）	—$CH_2CH_2CH_2CH_2CH_3$
Isopentyl（异戊基）	—$CH_2CH_2CH(CH_3)_2$
Hexyl（己基）	—$CH_2CH_2CH_2CH_2CH_2CH_3$
Heptyl（庚基）	—$CH_2CH_2CH_2CH_2CH_2CH_2CH_3$
Octyl（辛基）	—$CH_2CH_2CH_2CH_2CH_2CH_2CH_2CH_3$

Table 3.2 General formula for selected classes of organic compounds

Class	General formula	Functional group
Alcohol（醇）	ROH	Hydroxyl group（羟基）
Aldehyde（醛）	RC(O)H	Carbonyl group（羰基）
Ketone（酮）	RC(O)R′	Carbonyl group（羰基）
Carboxylic acid（羧酸）	RC(O)OH	Carboxyl group（羧酸基）
Ester（酯）	RC(O)OR′	Ester group（酯基）
Amine（胺）	RNH_2/RNR′H/RNR′R″	Amino group（胺基）
Amide（酰胺）	RC(O)NH_2 RC(O)NR′H RC(O)NR′R″	Amide group（酰胺基）

The functional groups in some selected organic compounds shown in Table 3.2 are the **hydroxyl** group (in **alcohol**), the carbonyl group (in **aldehydes** and **ketones**), the carboxyl group (in **carboxylic acids**), the —C(O)OR group (in **esters**), the amino group (in **amines**), and the amide group (in **amides**). Functional groups determine many of the properties of a compound, including what types of reactions it may undergo.

An alcohol contains an alkyl group R and the functional group —OH. The identity of an individual alcohol depends on the identity of the alkyl group R. For example, when R is the methyl group, it is CH_3OH. This is methanol or methyl alcohol, also known as wood alcohol. It is highly toxic and can cause blindness or even death in relatively small doses. When R is the ethyl group, it is CH_3CH_2OH. This is ethanol or ethyl alcohol. Ethanol is the alcohol in alcoholic beverages. When R is the isopropyl group, it is $(CH_3)_2CHOH$. This is isopropyl alcohol. Isopropyl alcohol, what we commonly call "rubbing alcohol", is widely used as a disinfectant.

Aldehyde and ketone contain the carbonyl group with the formula RC(O)R. If one of the R group is a H atom, the compound is called aldehyde. If both R groups is alkyl or aromatic (aryl) group, the compound is called ketone.

A carboxylic acid, RC(O)OH, is a compound consisting of an alkyl group (or aromatic group) R and the functional group carboxyl group —C(O)OH (carbonyl and hydroxyl). If R is an aliphatic residue, these compounds are called fatty acid since high-molecular weight compounds of this type are readily got from naturally occurring fats and oils.

An ester, RC(O)OR′, is the product from the reaction of a carboxylic acid, RC(O)OH, and an alcohol, R′OH. This type of reaction is called a condensation reaction, in which two molecules are combined by the elimination of a molecule such as water.

Amines are organic derivatives of ammonia (NH_3), in which one or more organic groups (R) are substituted for H atoms. Their classification is based on the number of R groups bonded to the nitrogen atom—one for primary amines, two for secondary amines, and three for tertiary amines.

The functional group of an amide is an acyl group, RC(O), bonded to a nitrogen atom. Like amines, primary, secondary, and tertiary amides are classified by the number of R groups bonded to the nitrogen atom (one, two, and three, respectively).

Many compounds contain more than one functional group. For example, amino acids contain both the amino group and the carboxyl group.

New Words and Expressions

acetone /ˈæsɪtəʊn/ n. 丙酮
alkane /ˈælkeɪn/ n. 烷烃
alkyl group n. 烷基
amino acid n. 氨基酸
disinfectant /ˌdɪsɪnˈfektənt/ n. 消毒剂

functional group n. 官能团
methane /ˈmiːθeɪn/ n. 甲烷
multiple bond n. 多重键
propanol /ˈprəʊpənɒl/ n. 丙醇
rubbing alcohol n. 医用酒精

Questions

Give some examples of functional groups in organic compounds.

3.3 Nomenclature of Organic Compounds

There are millions of organic compounds found in the world. To bring order to the naming of newly discovered compounds, the International Union of Pure and Applied Chemistry (IUPAC) established formal rules for naming organic compounds. The following is some fundamental rules for naming organic compounds.

3.3.1 Alkane and Cycloalkane

The name of an alkane is based on the number of carbon atoms in the longest chain in the molecule. Specific names are used to indicate the number of carbon atoms in the chain, as shown in Table 3.3.

Table 3.3 Alkane names for chain lengths up to carbons

Number of carbon atoms	Name	Molecular formula
1	Methane (甲烷)	CH_4
2	Ethane (乙烷)	C_2H_6
3	Propane (丙烷)	C_3H_8
4	Butane (丁烷)	C_4H_{10}
5	Pentane (戊烷)	C_5H_{12}
6	Hexane (己烷)	C_6H_{14}
7	Heptane (庚烷)	C_7H_{16}
8	Octane (辛烷)	C_8H_{18}
9	Nonane (壬烷)	C_9H_{20}
10	Decane (癸烷)	$C_{10}H_{22}$

A fragment of a carbon chain that is missing a single hydrogen is named by replacing the -*ane* ending of the carbon chain to -*yl*. For example, the two-carbon fragment CH_3CH_2- is the ethyl group.

The alkanes in Table 3.3 are examples of unsaturated hydrocarbons, organic compounds where each carbon is bonded to no more than two other carbon atoms. The name of an unbranched hydrocarbon is based on the number of carbon atoms in the molecule, as shown in the table. In branched hydrocarbons, at least one carbon is connected to more than two carbon atoms, and the naming is more complex. The name of a branched hydrocarbon consists of a parent name, which indicates the longest carbon chain in the molecule, and substituent names, which indicate the fragments that are attached to the longest chain.

Rules for naming branched alkanes and cycloalkanes are listed in the following:

(1) Determine the longest parent name for the compound by identifying the number of carbon atoms in the longest continuous carbon chain in the molecule. For a cycloalkane, the number of atoms in the ring gives the parent name.

(2) Name any alkane substituents attached to the chain by dropping the -*ane* from the alkane name and adding -*yl*.

(3) Number the longest continuous carbon chain to place substituents on carbon atoms with the lowest possible number. For cycloalkanes, number the ring to locate the substituents at the lowest possible numbers, with substituents lowest in alphabetical order at lowest numbered carbons.

(4) Name the substituents first, in alphabetical order. Use numbers to indicate the location of the branching and prefixes (*di-*, *tri-*, *tetra-*, etc.) to indicate multiple identical substituents.

(5) Follow the substituent names with the parent name of the longest carbon chain.

Let us consider the alkanes having five carbons. The first compound is named pentane because it is a five-carbon unbranched hydrocarbon. In the second structure, the longest continuous carbon chain contains four carbons, so it has the parent name butane. There is a methyl group, $-CH_3$, on the second carbon in the chain, so the full name of the compound is 2-methylbutane. The third compound has three carbons (propane) in its longest chain and two methyl substituents on the second carbon in the three-carbon chain. Therefore, the compound name is 2,2-dimethylpropane.

pentane   2-methylbutane   2,2-dimethylpropane

Cycloalkane are also named using the largest number of continuously connected carbon atoms. An unbranched cycloalkane is named using the prefix *cyclo-*, and a branched cycloalkane is named using the same rules.

cyclopantane   cyclobutane   1-ethyl-2-methylcyclopentane

3.3.2 Alkenes

Alkenes are hydrocarbons that contain at least one carbon-carbon double bond. Alkenes with C=C and no rings have the general formula C_nH_{2n}. Alkene names contain a parent alkane name with the *-ane* ending changed to *-ene*. In larger alkanes a numbering system is used to indicate the position of the double bond. Branched alkenes are named using the same rules used for branched alkanes.

$H_2C=CH-CH_2-CH_3$ $H_3C-CH=CH-CH_3$
1-butene   2-butene

$H_2C=C(CH_3)-CH_2-CH_3$ $H_3C-C(CH_3)=CH-CH_3$
2-methyl-1-butene   2-methyl-2-butene

3.3.3 Alkynes

Alkynes are hydrocarbons that contain at least one C≡C. Alkynes with one carbon-carbon triple bond and no rings have the general formula C_nH_{2n-2}. Like alkenes, alkynes are named using the parent alkane name with the *-ane* ending changed to *-yne*. The simplest alkyne, C_2H_2, is commonly named acetylene, but its IUPAC name is ethyne. The position of the triple bond along the chain is specified by number in a manner analogous to alkene nomenclature.

HC≡CCH₃   HC≡CCH₂CH₃   CH₃C≡CCH₃   (CH₃)₃CC≡CCH₃
propyne   1-butyne   2-butyne   4,4-dimethyl-2-pentyne

3.3.4 Arenes

Arenes, also called aromatic hydrocarbons, are compounds that contain a benzene ring, a six-member carbon ring of alternating single and double carbon-carbon bonds. Benzene, toluene, and naphthalene are some examples of arenes.

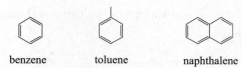

benzene   toluene   naphthalene

3.3.5 Alkyl halides

There are two different ways to name alkyl halides by IUPAC. In functional class nomenclature the alkyl group and the halide (fluoride, chloride, bromide, or iodide) are designated as separate words. The alkyl group is named on the basis of its longest continuous chain beginning at the carbon to which the halogen is attached.

$$CH_3Br \qquad CH_3CH_2CH_2CH_2CH_2F$$

methyl bromide   pentyl fluoride

Substitutive nomenclature of alkyl halides treats the halogen as a halo (*fluoro-*, *chloro-*, *bromo-*, or *iodo-*) substituent on an alkane chain. The carbon chain is numbered in the direction that gives the substituted carbon the lower number.

$$CH_3CH_2CH_2CH_2CH_2Br$$

1-bromopentane   1-iodocyclohexane

When the carbon chain bears both a halogen and an alkyl substituent, the two are considered of equal rank, and the chain is numbered so as to give the lower number to the substituent nearer the end of the atom.

5-chloro-2-methylheptane   2-chloro-5-methylheptane

3.3.6 Alcohols

Functional class names of alcohols are derived by naming the alkyl group that bears the hydroxyl substituent (—OH) and then adding *alcohol* as a separate word. The chain is always numbered beginning at the carbon to which the hydroxyl group is attached.

Substitutive names of alcohols are developed by identifying the longest continuous chain that bears the hydroxyl group and replacing the *-e* ending of the corresponding alkane by the suffix *-ol*. The position of the hydroxyl group is indicated by number choosing the sequence that assigns the lower locant to the carbon that bears the hydroxyl group.

Functional class name	ethyl alcohol	1-methylpentyl alcohol
Substitutive name	ethanol	2-hexanol

3.3.7 Aldehydes

Identify the longest continuous carbon chain that includes the carbonyl group, name it according to the number of carbons it contains, and change the -e ending to -al. Number the C atoms starting with the carbonyl carbon. Use numbers and prefixes to indicate the position and the identity of any substituents.

ethanal (acetaldehyde)   propanal (propionaldehyde)   butanal   5-methylhexanal

3.3.8 Ketones

Identify the longest continuous carbon chain that includes the carbonyl group. Name it according to the number of carbons it contains, and change the -e ending to -one. If necessary, number the C atoms to give the carbonyl carbon the lowest possible number. Use numbers and prefixes to indicate the position and the identity of any substituents.

propanone (acetone)   2-butanone (ethyl methyl ketone)   5-methyl-3-hexanone

3.3.9 Amines

An amine with one carbon attached to nitrogen is a **primary amine**, an amine with two is a **secondary amine**, and an amine with three is a **tertiary amine**. The group attached to nitrogen may be any combination of alkyl or aryl groups.

Amines are named in two main ways in the IUPAC system, either as alkylamines or as alkanamines. When primary amines are named as alkylamines, the ending -amine is added to the name of the alkyl group that bears the nitrogen. When named as alkanamines, the alkyl group is named as an alkane and the -e ending replaced by -amine.

ethylamine (ethanamine)   cyclohexylamine (cyclohexanamine)   1-methylbutylamine (2-pentanamine)

Aniline is the parent IUPAC name for amino-substituted derivatives of benzene. Substituted derivatives of aniline are numbered beginning at the carbon that bears the amino group. Substituents are listed in alphabetical order, and the direction of numbering is governed by the usual "first point of difference" rule.

p-fluoroaniline or 4-fluoroaniline   5-bromo-2-ethylaniline

Arylamines may also be named as arenamines. e. g. Benzenamine is an alternative, but rarely used, name for aniline.

Compounds with two amino groups: suffix: *-diamine* to the name of the corresponding alkane or arene. The final *-e* of the parent hydrocarbon is retained.

1,2-propanediamine   1,6-pentanediamine   1,4-benzenediamine

Secondary and tertiary amines are named as *N*-substituted derivatives of primary amines. The parent primary amine is taken to be the one with the longest carbon chain. The prefix *N-* is added as a locant to identify substituents on the amine nitrogen.

N-methylethylamine   4-chloro-*N*-ethyl-3-nitroaniline   *N,N*-dimethylcycloheptylamine

3.3.10 Ether

Ether are named, in substitutive IUPAC nomenclature, as alkoxy derivatives of alkanes. First, select the name of the longer carbon chain joined to the oxygen and give it the alkane name. The remaining alkoxy group (the RO— group) is named as a functional group bonded to the alkane. Thus, $CH_3O\text{-}CH_2CH_2CH_3$ is named methoxypropane.

Functional class IUPAC names of ethers are derived by listing the two alkyl groups in the general structure ROR' in alphabetical order as separate words, and then adding the word *ether* at the end. For example, methyl *tert*-butyl ether and diethyl ether have the structural formulas, respectively.

tert-butyl ether   diethyl ether

3.3.11 Sulfide

The sulfur analogs (RS—) of alkoxy groups are called *alkylthio* groups. The following examples illustrate the use of alkylthio prefixes in substitutive nomenclature of sulfides. Functional class IUPAC names of sulfides are derived in exactly the same way as those of ethers but end in the word *sulfide*.

	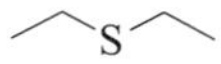	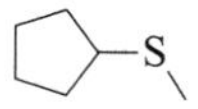
Substitutive name	ethylthioethane	(methylthio)cyclopentane
Functional class name	diethyl sulfide	cyclopentyl methyl sulfide

3.3.12 Carboxylic acids

Identify the longest continuous carbon chain that includes the carboxy group. Name it according to the number of carbons it contains, and change the -e ending to -oic acid. Number the C atoms starting with the carbonyl carbon. Use numbers and prefixes to indicate the position and the identity of any substituents.

Many organic compounds have common names in addition to their systematic names. Common names for some of the carboxylic acids shown here are given in parentheses.

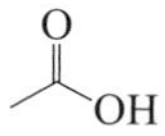

ethanoic acid (acetic acid)　　propanoic acid (propionic acid)　　butanoic acid (butyric acid)　　5-methylhexanoic acid

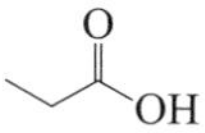

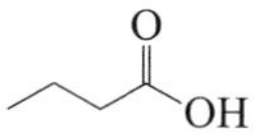

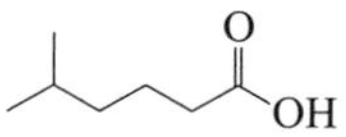

3.3.13 Acyl halide

With the exception of nitriles (RC≡N), all carboxylic acid derivatives consist of an acyl group [RC(O)-] attached to an electronegative atom. Acyl group are named by replacing the -ic acid ending of the IUPAC name of the corresponding carboxylic acid by -yl.

Acyl halide are named by replacing the name of the appropriate halide after that of the acyl group.

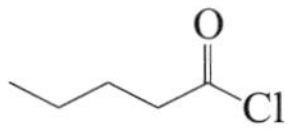

pentanoyl chloride 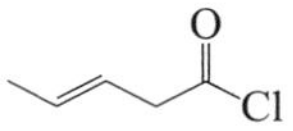　　3-pentenoyl chloride 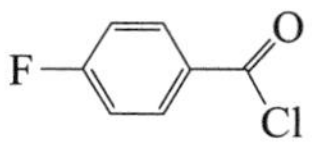　　p-fluorobenzoyl chloride

3.3.14 Acid Anhydrides

In naming **carboxylic acid anhydrides** in which both acyl groups are the same, we simply specify the acid and replace acid by *anhydride*. When the acyl groups are different, they are cited in alphabetical order.

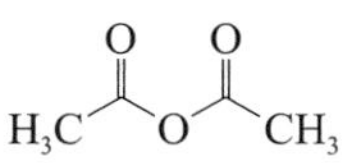

acetic anhydride 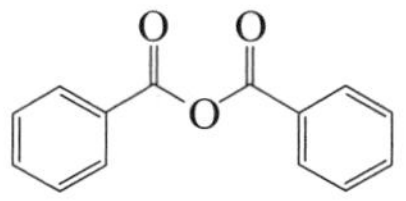　　benzoic anhydride 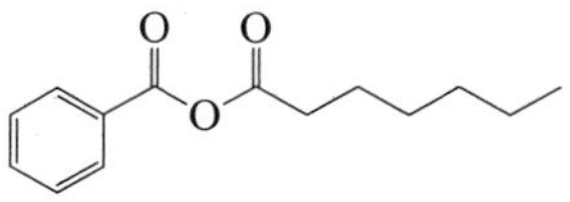　　benzoic heptanoic anhydride

3.3.15 Esters

Esters are named as **alkyl alkanoates**. The alkyl group R′ or RC(O)OR′ is cited first, followed by the acyl portion RC(O)-. The acyl portion is named by substituting the suffix *-ate* for the *-ic acid* ending of the corresponding acid.

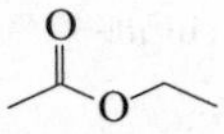

ethyl ethanoate
(ethyl acetate)

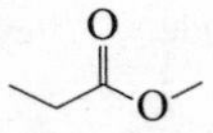

methyl propanoate

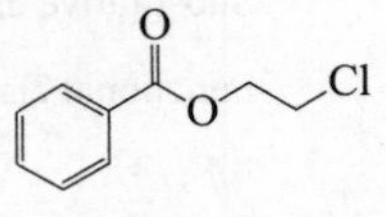

2-chloroethyl benzoate

3.3.16 Amides

The names of amides of the type $RC(O)NH_2$ are derived from carboxylic acids by replacing the suffix *-oic acid* or *-ic acid* by *-amide*.

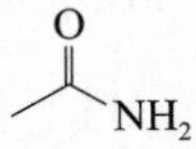

acetamide

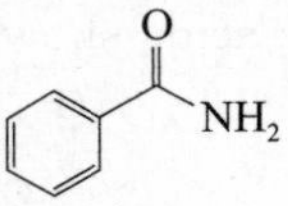

benzamide

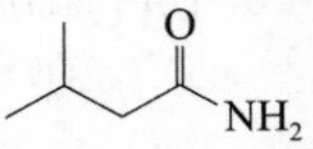

3-methylbutanamide

We name compounds of the type $RC(O)NHR'$ and $RC(O)NR'_2$ as *N*-alkyl and *N*,*N*-dialkyl substituted derivatives of a parent amide.

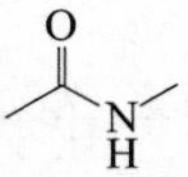

N-methylacetamide

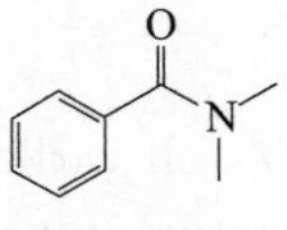

N,*N*-diethylbenzamide

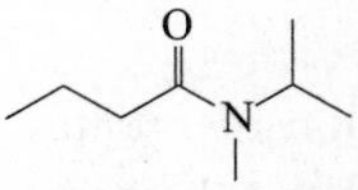

N-isopropyl-*N*-methylbutanamide

3.3.17 Nitriles

Substitutive IUPAC names for nitriles add the suffix *-nitrile* to the name of the parent hydrocarbon chain that includes the carbon of the cyano group. Nitriles may also be named by replacing the *-ic acid* or *-oic acid* ending of the corresponding carboxylic acid with *-onitrile*. Alternatively, they are sometimes given functional class IUPAC names as alkyl cyanides.

ethanenitrile
acetonitrile

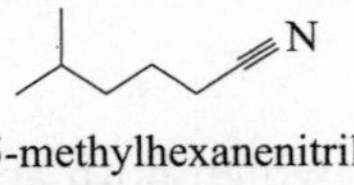

5-methylhexanenitrile
4-methylpentyl cyanide

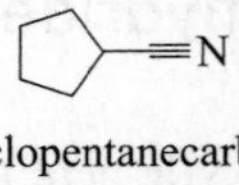

cyclopentanecarbonitrile
cyclopentyl cyanide

New Words and Expressions

branched hydrocarbon n. 支链烃
chain /tʃeɪn/ n. 链
derivative /dɪˈrɪvətɪv/ n. 衍生物
double bond n. 双键
fragment /ˈfræɡmənt/ n. 片段

parent name n. 母体名
primary amine n. 一级胺
saturated hydrocarbon n. 饱和烃
secondary amine n. 二级胺
tertiary amine n. 三级胺

triple bond n. 三键　　　　　　　　unsaturated hydrocarbon n. 不饱和烃

Questions

Write the names for the following compounds:

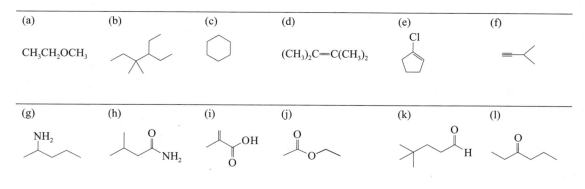

3.4 Hydrocarbons

The simplest organic compounds are **hydrocarbons**, *compounds containing only carbon and hydrogen*. All other organic compounds, for example, those containing O, N, and the halogen atoms are classified as being derived from hydrocarbons. Due to the simple composition of hydrocarbons, you might think that they represent a very limited set of molecules. However, several hundred thousand hydrocarbons were found or synthesized.

Hydrocarbons can be classified into three main groups:

(1) **Saturated hydrocarbons** are *hydrocarbons that contain only carbon-carbon single bonds*. Saturated hydrocarbon molecules can be cyclic or acyclic. A cyclic hydrocarbon is one in which a chain of carbon atoms has formed a ring. An acyclic hydrocarbon is one that does not contain a ring or carbon atoms.

(2) **Unsaturated hydrocarbons** are *hydrocarbons that contain at least one carbon-carbon double or triple bond*.

(3) **Aromatic hydrocarbons** are *hydrocarbons that contain benzene rings or similar features*.

The saturated and unsaturated hydrocarbons are often referred to as the aliphatic hydrocarbons.

3.4.1 Alkanes and Cycloalkanes

The **alkanes** are *acyclic saturated hydrocarbons*, and the **cycloalkanes** are *cyclic* *saturated hydrocarbons*. The simplest hydrocarbon, an alkane called methane, consists of one carbon atom to which four hydrogen atoms are bonded in a tetrahedral arrangement. You can represent methane by its molecular formula, CH_4, which gives the number and kind of atoms in the molecule, or by its structural formula, which shows how the atoms are bonded to one another.

$$CH_4 \qquad\qquad \begin{array}{c} H \\ | \\ H-C-H \\ | \\ H \end{array}$$

molecular formula of methane   structural formula of methane

Note that the structural formula does not convey information about the three-dimensional arrangement of the atoms. You may use ball-and-stick model or spacing-filling model to draw the three-dimensional formula depicting the molecular geometry.

Methane is a very important molecule since it is the principal component of natural gas. In 2018, more than 2.8×10^{11} m³ natural gas was consumed in China to supply heating, power generation, transportation, and industrial needs.

3.4.1.1 *The Alkane Series*

The alkane, also called **paraffins** (coming from the Latin *parum affinus*, meaning "little affinity"), have the general formula C_nH_{2n+2}. For $n=1$, methane, the formula is CH_4; for $n=2$, C_2H_6; $n=3$, C_3H_8, and so on. Note that the general formula conveys no information about how the atoms are connected. Now, we assume that the carbon atoms are bonded together in a straight chain with hydrogen atoms completing the four required bonds to each carbon atom: these are called *straight-chain* or *normal* alkanes. The structural formulas for the first four straight chain alkanes are shown.

methane   ethane   propane   butane

Because carbon atoms typically have four bonds, we often write the structures of the parts of organic compounds using **condensed structure formulas**, or **condensed formulas**, *where the bonds around each carbon atom in the compound are not explicitly written*. For example,

$\begin{array}{c} H \\ | \\ H-C- \\ | \\ H \end{array}$ is written as CH_3 $\begin{array}{c} H \\ | \\ -C- \\ | \\ H \end{array}$ is written as CH_2

$\begin{array}{c} H \\ | \\ -C- \\ | \end{array}$ is written as CH $\begin{array}{c} | \\ -C- \\ | \end{array}$ is written as C

Condensed formulas of the first four alkanes ($n=1\sim 4$) are

     CH_4 CH_3CH_3 $CH_3CH_2CH_3$ $CH_3CH_2CH_2CH_3$

     methane   ethane   propane   butane

Note that the condensed formula of an alkane differs from that of the preceding alkane ($n-$

1) by a CH_2 group. These alkanes constitute a **homologous series**, which is *a series of compounds in which one compound differs from a preceding one by a fixed group of atoms*. Members of a homologous series have similar chemical properties, and their physical properties change throughout the series in a regular way. Table 3.4 lists the melting points and boiling points of the first ten straight-chain alkanes ($n=1 \sim 10$). We can find that the melting points and boiling points generally increase in the series with an increase in the number of carbon atoms in the chain. This is a result of increasing intermolecular forces, which increase with molecular weight.

Table 3.4 Physical properties of straight-chain alkanes

Name	Number of carbons	Formula	(Melting point/Boiling point)/℃
Methane	1	CH_4	$-183/-162$
Ethane	2	CH_3CH_3	$-172/-89$
Propane	3	$CH_3CH_2CH_3$	$-187/-42$
Butane	4	$CH_3(CH_2)_2CH_3$	$-138/0$
Pentane	5	$CH_3(CH_2)_3CH_3$	$-130/36$
Hexane	6	$CH_3(CH_2)_4CH_3$	$-95/69$
Heptane	7	$CH_3(CH_2)_5CH_3$	$-91/98$
Octane	8	$CH_3(CH_2)_6CH_3$	$-57/126$
Nonane	9	$CH_3(CH_2)_7CH_3$	$-54/151$
Decane	10	$CH_3(CH_2)_8CH_3$	$-30/174$

3.4.1.2 *Constitutional Isomerism and Branched-Chain Alkanes*

In addition to the straight-chain alkanes, branched-chain alkanes are possible. For example, isobutane (or 2-methylpropane) has the structure:

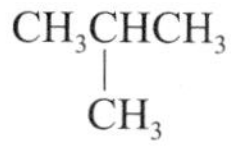

$$\begin{array}{c} CH_3CHCH_3 \\ | \\ CH_3 \end{array}$$

isobutane
(2-methylpropane)

Isobutane, C_4H_{10}, has the same molecular formula as butane, the straight-chain hydrocarbon. However, isobutane and butane have different structural formulas and, therefore, different molecular structure. Butane and isobutane are **constitutional** (or **structural**) **isomers**, *compounds with the same molecular formula but different structural formulas*. Because these isomers have different structures, they have different physical properties. For example, isobutane boils at $-12℃$, whereas butane boils at $0℃$. The difference in boiling point for these two isomers can be attributed to the fact that isobutane has a more compact molecular structure than butane, which results in weaker intermolecular interactions between isobutane molecules.

Branched alkanes are usually written using condensed structural formulas.

3.4.1.3 *Cycloalkanes*

The general formula for cyclic cycloalkanes is C_nH_{2n}. The first three members of the cycloalkane series along with their names and condensed structural formulas are shown below:

Name	cyclopropane	cyclobutane	cyclopentane
Molecular formula	C_3H_6	C_4H_8	C_5H_{10}
Condensed structural formula	△	□	⬠

In the condensed structural formulas, a carbon atom and its attached hydrogen atoms are assumed to be at each corner.

3.4.1.4 *Sources and Uses of Alkanes and Cycloalkanes*

Fossil fuels (natural gas, petroleum, and coal) are the principal sources of all types of organic chemicals. Natural gas is a mixture of low molecular-weight hydrocarbons made up primary of methane (CH_4), with lesser amounts of ethane (C_2H_6), propane (C_3H_8), and butane (C_4H_{10}). Petroleum, or crude oil, is the raw material extracted from a well. It is a mixture of alkanes and cycloalkanes with small amounts of aromatic hydrocarbons. The composition of petroleum is not consistent, it is dependent on geologic location and the organic matter present during oil formation. Because of this, a barrel of oil from the North Slope of Canada contains a much different mixture of organic materials than a barrel of oil from Saudi Arabia. The Saudi crude oil might be made up primary of molecules with 5~20 carbon atoms, whereas the Canada crude might consist mainly of molecules containing 20~40 carbon atoms. Crude oils that contain a majority of the low-molecular-weight hydrocarbons are often more desirable because they are easier to transport and require less refining to convert to high-demand products like gasoline. Current world oil consumption is in excess of 90 million barrels per day.

Because fossil fuels are extracted from their source as mixtures of hydrocarbons, it is usually necessary to separate these mixtures into various components. Such separations are most easily performed by distilling the mixture into fractions that contain mixtures of compounds of different molecular weight. Table 3.5 lists the fractions distilled from petroleum. Every fraction is the mixture of hydrocarbons.

Table 3.5 Fractions from the distillation of petroleum

Boiling range /℃	Name	Range of carbon atoms per molecule	Use
Below 20	Gases	C_1 to C_4	Heating, cooking, and petrochemical raw material
20~200	Naphtha; Straight-run gasoline	C_5 to C_{12}	Fuel; lighter fractions (such as petroleum ether, b.p. 30~60℃) are also used as laboratory solvents
200~300	Kerosene	C_{12} to C_{15}	Fuel
300~400	Fuel oil	C_{15} to C_{18}	Heating homes, diesel fuel
Over 400		Over C_{18}	Lubricating oil, greases, paraffin waxes, asphalt

Often, there is a need to further separate the fractions listed in Table 3.5 into molecules with the same molecular weight, molecular formula, or structure. Through chemical processes that usually involve catalysts, small molecules can be combined into larger molecules, and large molecules can be broken apart. The processing of petroleum via distillation or chemical reaction is called **petroleum refining**. The type and extent of the petroleum refining that is performed depend on the type of crude oil that is available and on the demand for a particular type or product. One chemical process, called **catalytic cracking**, involves passing hydrocarbon vapor over a heated catalyst of alumina (Al_2O_3) and silica (SiO_2) to break apart, or "crack", high-molecular-weight hydrocarbons to produce hydrocarbons of low molecular weight. This process can be used to convert the fuel oil fraction of petroleum to gasoline.

Petroleum refining also involves the conversion of the relatively abundant alkanes to unsaturated hydrocarbons, aromatic hydrocarbons, and hydrocarbon derivatives. The alkanes serve as the starting point for the majority of organic compounds, including plastics and pharmaceutical drugs. It is interesting that many materials we used today come from alkanes or cycloalkanes.

3.4.1.5 *Reactions of Alkanes with Oxygen*

You may think that alkanes are not particularly reactive molecules at normal temperature. However, in our lives, you would have to be extremely careful about how you filled your automobile with fuel, and you need to pay attention to using natural gas in your home for heating or cooking. In addition, the refining process could not involve the separation of the crude oil components at high temperature by distillation. In fact, we do react alkanes everyday through combustion with O_2; all hydrocarbon burn (combust) in an excess of O_2 at elevated temperature to produce carbon dioxide, water, and heat. For example, a propane gas grill uses the reaction:

$$C_3H_8(g) + 5O_2(g) \longrightarrow 3CO_2(g) + 4H_2O(l) \qquad \Delta H^\ominus = -2220 \text{ kJ/mol}$$

The large negative $\Delta H^\ominus$ value for this reaction and all hydrocarbon reactions with oxygen demonstrates why we rely on these molecules to meet our energy needs.

3.4.1.6 *Substitution Reactions of Alkanes*

Under the right conditions, alkanes can react with other molecules. An important example is the reaction of alkanes with the halogens F_2, Cl_2, and Br_2. Methane reacts with Cl_2 in the presence of light (induced by $h\nu$) or heat to produce methyl chloride (or chloromethane).

$$\underset{\underset{H}{|}}{\overset{\overset{H}{|}}{H-C-H}} + Cl-Cl \xrightarrow{h\nu} \underset{\underset{H}{|}}{\overset{\overset{H}{|}}{H-C-Cl}} + H-Cl$$

This is an example of a substitution reaction. A **substitution reaction** is *a reaction in which a part of the reacting molecule is substituted for an H atom on a hydrocarbon or hydrocarbon group*. All of the H atoms of an alkane may undergo substitution, leading to a mixture of products.

$$CH_3Cl + Cl_2 \xrightarrow{h\nu} CH_2Cl_2 + HCl$$
$$CH_2Cl_2 + Cl_2 \xrightarrow{h\nu} CHCl_3 + HCl$$
$$CHCl_3 + Cl_2 \xrightarrow{h\nu} CCl_4 + HCl$$

The CCl_4 product can be reacted with HF in the presence of a $SbCl_5$ catalyst to produce trichlorofluoromethane, CCl_3F, also known as CFC-11.

$$CCl_4 + HF \xrightarrow{SbCl_5} CCl_3F + HCl$$

This compound is one of a number of chlorofluorocarbons (CFCs) used for much of the twentieth century as a refrigerant. Data obtained in the 1970s revealed that these compounds survived long enough to travel to the stratospheric region of our atmosphere, where they facilitate the destruction of the ozone layer. Recent refrigerants that do not as readily contribute to the ozone destruction include hydrofluorocarbons (HFCs), such as CH_3CH_2F, which do not contain chlorine atoms. These HFCs are now used to replace CFCs.

3.4.2 Alkenes

Alkenes are unsaturated hydrocarbons, which carbon-carbon double bonds. They are typically much more reactive than alkanes.

Under the proper conditions, molecular hydrogen can be added to an alkane to produce a saturated compound in a process called **catalytic hydrogenation**. For example, ethylene adds hydrogen to give ethane.

$$\underset{\text{ethylene}}{H_2C=CH_2} + H_2 \xrightarrow{\text{Ni catalyst}} \underset{\text{ethane}}{H_3C-CH_3}$$

Catalytic hydrogenation is also used in the food industry to convert (hydrogenate) carbon-carbon double bonds to carbon-carbon single bonds. For example, margarine can be manufactured by hydrogenating some of the double bonds present in corn oil to change it from oil to solid (fat).

3.4.2.1 *Oxidation Reactions of Alkenes*

Because alkenes are hydrocarbons, they undergo complete combustion reactions with oxygen at high temperature to produce carbon dioxide and water. Unsaturated hydrocarbons can also be partially oxidized under relatively mile conditions. For example, when aqueous potassium permanganate, $KMnO_4(aq)$, is added to an alkene (or alkyne), the purple color of $KMnO_4$ fades and a brown precipitate of manganese dioxide forms.

$$3C_4H_9CH=CH_2 + 2MnO_4^-(aq) + 4H_2O \longrightarrow 3C_4H_9CH(OH)-CH_2(OH) + 2MnO_2(s) + 2OH^-(aq)$$

3.4.2.2 *Addition Reactions of Alkenes*

Alkenes are more reactive than alkanes because of the presence of the double bond. Many reactants add to the double bond. A simple example is the addition of a halogen, such as Br_2, to propene

$$CH_3CH=CH_2 + Br_2 \longrightarrow H_3C-\underset{\underset{Br}{|}}{\overset{\overset{H}{|}}{C}}-\underset{\underset{Br}{|}}{\overset{\overset{H}{|}}{C}}-H$$
propene

An **addition reaction** is *a reaction in which parts of a reactant are added to each carbon atom of a carbon-carbon double bond, which converts to a carbon-carbon single bond* (Addition to triple bonds is also possible, giving a product with a double bond). The addition of Br_2 to an alkene is fast. In fact, it occurs so readily that bromine dissolved in carbon tetrachloride, CCl_4, is a useful reagent to test for unsaturation. When a few drops of the solution are added to an alkene, the red-brown color of the Br_2 disappears immediately.

Unsymmetrical reagents, such as HCl and HBr, add to unsymmetrical alkenes to give two products that are constitutional isomers. For example,

$$CH_3\overset{3}{C}H=\overset{}{C}H_2 + HBr \longrightarrow H_3\overset{3}{C}-\overset{\overset{H}{|}}{\underset{\underset{Br}{|}}{\overset{2}{C}}}-\overset{\overset{H}{|}}{\underset{\underset{H}{|}}{\overset{1}{C}}}-H$$
2-bromopropane

and

$$CH_3\overset{3}{C}H=CH_2 + HBr \longrightarrow H_3\overset{3}{C}-\overset{\overset{H}{|}}{\underset{\underset{H}{|}}{\overset{2}{C}}}-\overset{\overset{H}{|}}{\underset{\underset{Br}{|}}{\overset{1}{C}}}-H$$
1-bromopropane

In one case, the hydrogen atom of HBr adds to carbon atom 1, giving 2-bromopropane; in the other case, the hydrogen atom of HBr adds to carbon atom 2, giving 1-bromopropane. (The name 1-bromopropane means that a bromine atom is substituted for a hydrogen atom at carbon atom 1.) However, the two products are not formed in equal amounts; one is more likely to form. **Markownikoff's rule** is *a generalization stating that the major product formed by addition of an unsymmetrical reagent such as H—Cl, H—Br, or H—OH is the one obtained when the H atom of the reagent adds to the carbon atom of the multiple bond that already has the greater number of hydrogen atoms attached to it.* In the preceding example, the H atom of HBr should add preferentially to carbon 1, which has two hydrogen atoms attached to it. The major product then is 2-bromopropane.

3.4.3 Alkynes

Alkynes are unsaturated hydrocarbons containing a carbon-carbon triple bond. The general formula is C_nH_{2n-2}. The simplest alkyne is acetylene (ethyne), $HC\equiv CH$, a linear molecule.

Acetylene is a very reactive gas that is used to produce a variety of other chemical compounds. It burns with oxygen in the oxyacetylene torch to give a very hot flame (about

3000℃). Acetylene is produced commercially from methane.

$$2CH_4 \xrightarrow{1600℃} HC{\equiv}CH + 3H_2$$

Acetylene is also prepared from calcium carbide, CaC_2. Calcium carbide is obtained by heating calcium oxide and coke (carbon) in an electric furnace

$$CaO(s) + 3C(s) \xrightarrow{2000℃} CaC_2(l) + CO(g)$$

The calcium carbide is cooled until it solidifies. The carbide ion, C_2^{2-}, is strongly basic and reacts with water to produce acetylene.

The alkynes, like alkenes, undergo addition reactions, usually adding two molecules of the reagent for each $C{\equiv}C$ bond. The major product is the isomer predicted by Markovnikov's rule, as shown in the following reaction

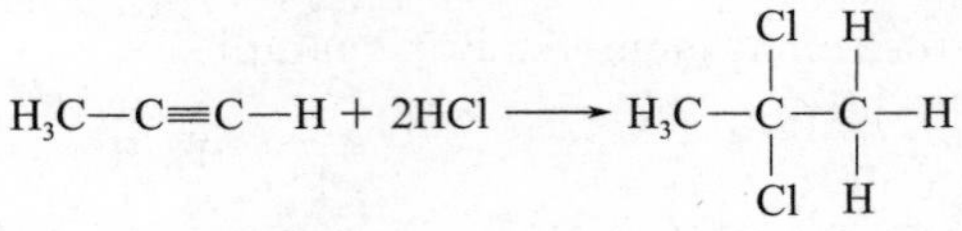

$$H_3C-C{\equiv}C-H + 2HCl \longrightarrow H_3C-\underset{\underset{Cl}{|}}{\overset{\overset{Cl}{|}}{C}}-\underset{\underset{H}{|}}{\overset{\overset{H}{|}}{C}}-H$$

3.4.4 Aromatic Hydrocarbons

Aromatic hydrocarbons usually contain benzene rings: six-membered rings of carbon atoms with alternating carbon-carbon single and carbon-carbon double bonds. The electronic structure of benzene can be represented by resonance formulas. For benzene

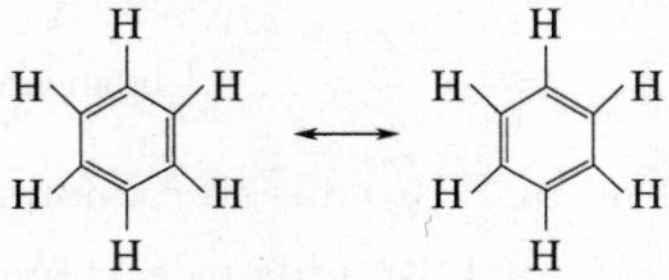

This electronic structure can also be described using molecular orbitals. In this description, π molecular orbitals encompass the entire carbon-atom ring, and the π electrons are said to be delocalized. Delocalization of electrons means that the double bonds in benzene do not behave as isolated double bonds. Two condensed formulas for benzene are

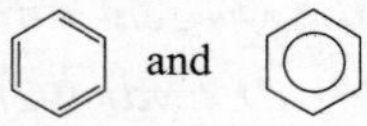

where the circle in the formula at right represents the delocalization of the π electrons (and therefore the double bonds). Although you will often encounter benzene represented with alternating double and single bonds as shown on the left, the better representation is the one that indicates the bond delocalization.

Aromatic compounds are found everywhere. The term **aromatic** implies that compounds that contain a benzene ring have aromas, and this is indeed the case. Flavoring agents that can be synthesized in the laboratory or found in nature include the flavor and aroma of cinnamon, cinnamaldehyde, and the wintergreen flavor of candies and gum, methyl salicylate.

Benzene rings also exist in the pain relievers acetylsalicylic acid (Aspirin) and acetaminophen (Tylenol).

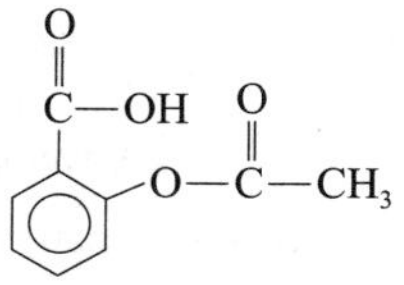

acetylsalicylic acid
(aspirin)

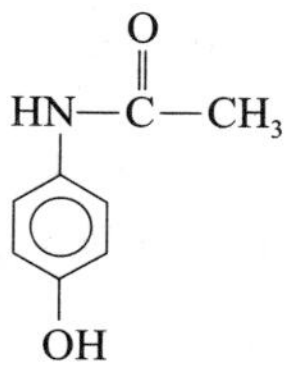

acetaminophen
(tylenol)

Other examples include the active ingredients in some sunscreens: oxybenzone and *p*-aminobenzoic acid (often abbreviated as PABA on the label).

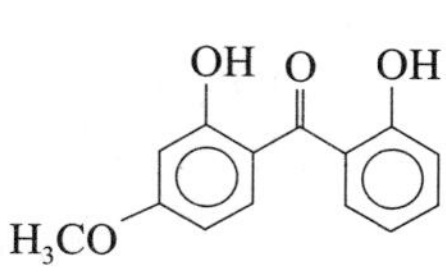

oxybenzone 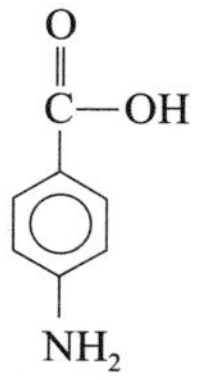   *p*-aminobenzoic acid (PABA)

Benzene rings can also fuse together polycyclic aromatic hydrocarbons in which two or more rings share carbon atoms. Naphthalene, anthracene, and phenanthrene are some examples of polycyclic aromatic hydrocarbons. Naphthalene is one of the compounds that gives mothballs their characteristic odor.

naphthalene   anthracene   phenanthrene

3.4.4.1 Characteristics of Aromatic Hydrocarbons

Aromatic hydrocarbons are highly flammable and should always be handled with care. Prolonged inhalation of benzene vapor results in a decreased production of both red blood cells and white blood cells, which can be fatal. In addition, benzene is a carcinogen. Benzene and some other toxic aromatic compounds have been isolated in the tar formed by burning cigarettes, in polluted air, and as a decomposition product of grease in the charcoal grilling of meat.

From a close examination of the structures of aromatic molecules, they all share two common features.

(1) They are planar (flat), cyclic molecules.

(2) They have a conjugated bonding system-a bonding scheme among the ring atoms that consists of alternating single and double bonds. The system must extend throughout the ring, and the π electron clouds associated with the double bonds must involve $(4n+2)$ electrons, where $n=1, 2, \cdots$, Thus, the benzene molecule has six electrons in the π electron clouds: $(4 \times 1)+2=6$. The naphthalene molecule has 14 $[(4 \times 3)+2=14]$. Neither of the two molecules depicted below is aromatic; both are aliphatic.

$$CH_2=CH-CH=CH-CH=CH_2$$

1,3,5-hexatriene   1,3-cyclopentadiene

The 1,3,5-hexatriene molecule has six π electrons in its conjugated bonding system, but it is not cyclic. The 1,3-cyclopentadiene molecule is cyclic but has only four π electrons in a conjugated bonding system that does not extend completely around the ring.

Benzene and its homologues are similar to other hydrocarbon in being insoluble in water but soluble in organic solvents. The boiling points of the aromatic hydrocarbons are slightly higher than those of the alkanes of similar carbon content. For example, hexane, C_6H_{14}, boils at 69℃, whereas benzene boils at 80℃. This can be explained by the planar structure and delocalized electron charge density of benzene, which increases the attractive forces between molecules. The symmetrical structure of benzene permits closer packing of molecules in the crystalline state and results in a higher melting point than for hexane. Benzene melts at 5.5℃ and hexane melts at -95℃.

When one of the six equivalent H atoms of a benzene molecule is removed, the resulting species is called a **phenyl group**. Two phenyl groups may bond together, as in biphenyl, or phenyl groups may be substituents in other molecules, as in phenylhydrazine, used in the detection of sugars.

phenyl group   biphenyl   phenylhydrazine

3.4.4.2 Uses of Aromatic Hydrocarbons

Current annual production of benzene in China is about 8 million tons. Over 90% of this is produced from petroleum. The process involves dehydrogenation and cyclization of hexane to the aromatic hydrocarbon. The most important use of the petroleum-produced benzene is in manufacturing ethylbenzene for the production of styrene plastics. Other applications include the manufacture of phenol, the synthesis of dodecylbenzene (for detergents), and as an octane enhancer in gasoline. The production of aromatic compounds by dehydration (removal of hydrogen) of alkanes yields large amount of hydrogen gas, which is an important reactant in the synthesis of ammonia.

3.4.4.3 Aromatic Substitution Reactions

Unlike alkenes and alkynes, aromatic molecules react by substitution, not by addition of atoms across a double bond. This difference in reactivity is due to the special stability of the electrons in the aromatic ring. In aromatic rings, certain electrons are delocalized into a π electron cloud. In alkene and alkyne molecules, the regions of high electron density associated with multiple bonds are localized between specific C atoms. These localized π bonds are much more reactive than the delocalized π bonds of the aromatic ring.

The substitution of a single atom or group, X, for a H atom in C_6H_6, can occur at any one of the six positions of the benzene ring. We say that the six positions are equivalent. If a group Y is substituted for a H atom in C_6H_5X, this question arises: to which of the remaining five positions does the Y group go? If all the sites on the benzene ring were equally preferred, the distribution of the products would be a purely statistical one. That is, there are five possible positions where Y can be substituted, and we should get 20% of each one. Since there are two possibilities that lead to an *ortho* isomer and two that lead to a *meta* isomer, however, we should expect the distribution of products to get 40% *ortho*, 40% *meta*, and 20% *para*.

The following scheme describes the products resulting from nitration followed by chlorination and chlorination followed by nitration. It shows that the substitution is not random. The —NO_2 group directs Cl to a *meta* position. Almost no *ortho* or para isomer is formed in reaction. The Cl group, on the other hand, is an *ortho*, *para* director. Essentially no *meta* isomer is produced in the reaction.

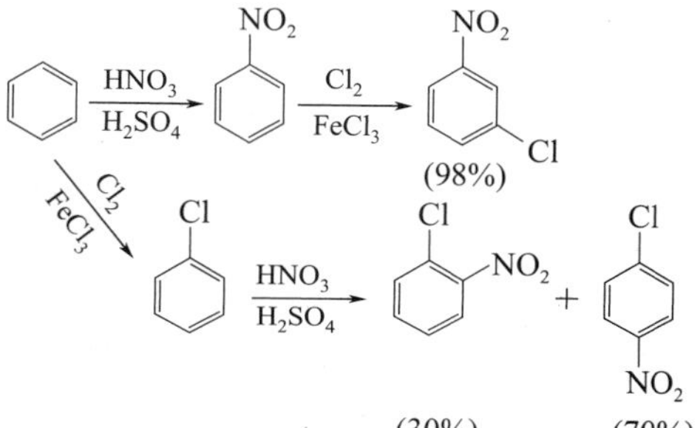

Whether a group is an *ortho*, *para*, or a *meta* director depends on how the presence of one substituent alters the electron distribution in the benzene ring. As a result, attack by a second group is more likely at one type of position than another. Examination of many reactions leads to the following order (from strongest to weakest).

Ortho, para directors: —NH_2, —OR, —OH, —OCOR, —R, —X (X=halogen)
Meta directors: —NO_2, —CN, —SO_3H, —CHO, —COR, —COOH, —COOR

When two groups of the same type (both *o*-, *p*- or both *m*-directing) are present, the stronger director wins out. When two groups of different type (one *o*-, *p*- and one *m*-directing) then it has been found that the *o*-, *p*-directing group guides the reaction.

New Words and Expressions

acetaminophen /əˌsiːtəˈmɪnəfen/ n. 对乙酰氨基酚
acetylsalicylic acid n. 乙酰水杨酸
acyclic /ˌeɪˈsaɪklɪk/ adj. 非环状的
addition reaction n. 加成反应
aliphatic hydrocarbon n. 脂肪烃
anthracene /ˈænθrəsiːn/ n. 蒽
aromatic hydrocarbon n. 芳烃

aspirin /ˈæsprɪn; ˈæspərɪn/ n. 阿司匹林
ball-and-stick model n. 球棍模型
barrel /ˈbærəl/ n. 桶
benzene ring n. 苯环
biphenyl /baɪˈfenəl/ n. 联(二)苯
catalytic cracking n. 催化裂化
catalytic hydrogenation n. 催化氢化
chlorination /ˌklɔːrɪˈneɪʃn/ n. 氯化

cinnamaldehyde /sɪnəˈmɔːldɪhaɪd/ n. 肉桂醛
cinnamon /ˈsɪnəmən/ n. 肉桂
constitutional (or structural) isomers n. 结构异构体
convey /kənˈveɪ/ v. 表达
delocalize /diːˈləʊk(ə)laɪz/ v. 离域
geologic /ˌdʒiːəˈlɒdʒɪk/ adj. 地质的
homologous series n. 同系列
Markownikoff's rule n. 马科尼科夫规则
methyl salicylate n. 水杨酸甲酯
naphthalene /ˈnæfθəliːn/ n. 萘
nitration /naɪˈtreɪʃən/ n. 硝化
ozone layer n. 臭氧层

p-aminobenzoic acid n. 对氨基苯甲酸
paraffin /ˈpærəfɪn/ n. 石蜡
petroleum refining n. 石油精炼
phenanthrene /fɪˈnænθriːn/ n. 菲
phenyl group n. 苯基
phenylhydrazine /ˌfiːnɪlˈhaɪdrəziːn/ n. 苯肼
spacing-filling model n. 空间充填模型
straight-chain n. 直链
stratospheric region n. 平流层
substitution reaction n. 取代反应
Tylenol /ˈtaɪlənɒl/ n. 泰诺（药品注册商标名）

Notes

1. The simplest hydrocarbon, an alkane called methane, consists of one carbon atom to which four hydrogen atoms are bonded in a tetrahedral arrangement.

参考译文：最简单的碳氢化合物是一种称为甲烷的烷烃类化合物，它由一个中心碳原子和四个氢原子以四面体形式排列结合而成。

2. One chemical process, called catalytic cracking, involves passing hydrocarbon vapor over a heated catalyst of alumina (Al_2O_3) and silica (SiO_2) to break apart, or "crack", high-molecular-weight hydrocarbons to produce hydrocarbons of low molecular weight.

参考译文：一种被称为催化裂化的化学过程是将碳氢化合物蒸气通过加热的氧化铝和二氧化硅催化剂，将高分子量的碳氢化合物分离或裂解，产生低分子量的碳氢化合物。

3. An addition reaction is a reaction in which parts of a reactant are added to each carbon atom of a carbon-carbon double bond, which converts to a carbon-carbon single bond.

参考译文：加成反应是将反应物加在具有碳-碳双键的碳原子中，从而转化为碳-碳单键的反应。

4. Markownikoff's rule is a generalization that states that the major product formed by addition of an unsymmetrical reagent such as H—Cl, H—Br, or H—OH H is the one obtained when the H atom of the reagent adds to the carbon atom of the multiple bond that already has the greater number of hydrogen atoms attached to it.

参考译文：马科尼科夫规则可概括为，通过加入不对称的试剂（如 H—Cl、H—Br 或 H—OH），其形成的主要产物是试剂的氢原子加在具有较多氢原子的不饱和碳原子上。

5. This can be explained by the planar structure and delocalized electron charge density of benzene, which increases the attractive forces between molecules.

参考译文：这可以用苯的平面结构和离域电子电荷密度来解释，因为它增加了分子间的吸引力。

> **Questions**

1. Give the condensed structural formulas of all possible substitution products of ethane and Cl_2.

2. Define the term *substitution reaction* and *addition reaction*. Give examples of each.

3.5 Nucleophilic Substitution Reactions

In this content we consider the essential features of substitution reactions involving compounds in which the functional group is bonded to a sp^3-hybridized carbon atom. We will focus primary on substitution reaction of haloalkanes. In a haloalkane, the carbon atom bonded to the halogen is sp^3-hybridized. With appropriate modifications, the concepts learned in this content can be used to understand substitution reaction of other compounds.

An example of substitution reaction involving a sp^3-hybridized carbon is the replacement of a Cl atom in chloromethane by a hydroxide group to give methanol:

$$CH_3Cl + OH^- \longrightarrow CH_3OH + Cl^-$$

This type of chemical reaction is called a **nucleophilic substitution reaction** because the hydroxide ions is a **nucleophile** (Nu:)—a reactant that seeks sites of low electron density in a molecule. The nucleophile is electron rich, and it is the species that donates a pair of electrons to another molecule. The C atom in CH_3Cl is considered electron deficient because the Cl atom draws electron density away from the carbon atom. Thus, the C atom is said to be electrophilic (electron attracting). The **electrophile** (E) is the species that is being attacked by the nucleophile and accepts a pair of electrons. Electrophiles contain atoms that are electron deficient. In the above reaction, the hydroxide ion attacks at the electron deficient C atom, displacing the chloride ion, which is called the **leaving group** (Lv).

Nucleophiles, often denoted by the abbreviation Nu, can be negatively charged or neutral, but every nucleophile contains at least one pair of unshared electrons. Nucleophiles are, in fact, Lewis bases. Here is the general equation for a nucleophilic substitution reaction.

$$Nu:^- + {-\overset{|}{\underset{|}{C}}-Lv} \longrightarrow {-\overset{|}{\underset{|}{C}}-Nu} + Lv^-$$

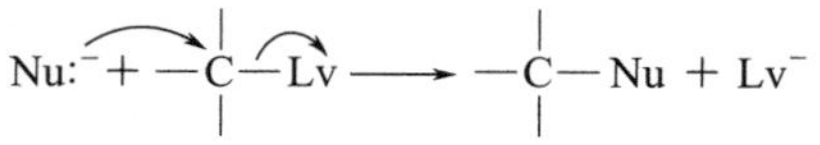

In a nucleophilic substitution reaction, the bond between carbon and the leaving group is broken and a new bond is formed by using a lone pair from the nucleophile. The electron pair from the bond that is broken ends up as a lone pair on the leaving group. In the viewpoints of acid-base reactions, equilibrium favors forming the substitution product if the leaving group is a weaker base than the nucleophile.

Nucleophilic substitution of haloalkanes occurs via one of two mechanisms. Before examining the mechanisms, let's consider a few specific examples of nucleophilic substitution reactions.

In the reaction shown below, the $CH_3C≡C^-$ ion is the nucleophile and CH_3CH_2Br is the electrophile:

$$H_3CC≡C:^- + H_3\underset{\beta}{CH_2}\underset{\alpha}{\overset{\delta^+}{C}}-\overset{\delta^-}{Br} \longrightarrow H_3CH_2C-C≡CCH_3 + Br^-$$

The α carbon is electron deficient (δ^+); it is the atom of the electrophile that is attacked by the nucleophile. Because both the nucleophile and the leaving group are Lewis bases, we can use acid-base concepts to establish that the reaction above will occur as written. The $CH_3C≡C^-$ ion has the negative charge localized on C, a very small atom compared with Br, and so the $CH_3C≡C^-$ is a much stronger base than Br^-. Because the reaction involves the formation of a weaker base, the reaction will occur as written.

Mechanisms of Nucleophilic Substitution Reactions: S_N1 and S_N2 are especially important.

In the nucleophilic substitution reaction, there are two types of mechanisms. The reaction between chloromethane and the hydroxide ion has a rate law that is first order in both the nucleophile and the electrophile. That is, rate=$k[OH^-][CH_3Cl]$. The mechanism for his reaction involves a bimolecular rate-determining step in which the nucleophilic group departs. The entering and leaving groups act at the same time in what is known as a concerted step, as depicted in this schematic representation:

$$HO:^- + H_3C-Cl \longrightarrow H_3C-OH + Cl^-$$

The arrow starts from an electron-rich center (the electrons on the hydroxy group in this case) to an electron-poor region (the C atom in the polar carbon-to-chlorine bond).

This mechanism is designated S_N2, with the S indicating substitution, the N indicating nucleophilic, and the 2 indicating that the rate-determining step is bimolecular. In the transition state, the HO^- ion attacks on the side opposite the Cl atom, and the C—Cl bond starts to break as the C—O bond simultaneously starts to form. In this mechanism, the nucleophile donates a pair of electrons to the electrophile to form a covalent bond.

Is there any experimental evidence to support this mechanism and postulated transition state? First, the rate law is suggestive of a bimolecular step; second, an observation by Paul Walden in 1893 confirmed the formation of the transition shown in Figure 3.1. He discovered that if the C atom bonded to the halogen is stereogenic, or chiral, the configuration at the chiral carbon is inverted—that is, the molecular structure is inverted and an enantiomer of the opposite configuration is formed. Thus, when (S)-2-iodobutane undergoes nucleophilic substitution by the hydroxide group, the compound (R)-2-butanol is formed (Figure 3.1). This inversion of configuration is taken as confirmation of the proposed bimolecular mechanism and the five-coordinate transition state.

The other mechanism of nucleophilic substitution at haloalkanes has a rate that is first order in the concentration of haloalkane only. For the reaction between 2-bromo-2-methylpropane and water, in which H_2O acts as the nucleophile

$$(CH_3)_3CBr + H_2O \longrightarrow (CH_3)_3COH + HBr$$

the rate law is: rate=$k[(CH_3)_3CBr]$.

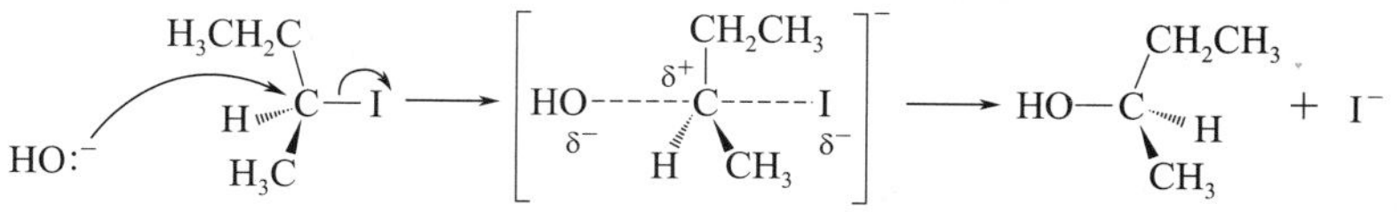

Figure 3.1 Inversion configuration in the S_N2 mechanism

This rate law suggests that the rate-determining step is unimolecular. The mechanism for this step is shown in Figure 3.2. The first step is a slow unimolecular step in which the haloalkane ionizes to form a bromide ion and a carbocation. The carbocation has a planar geometry and the positively charged carbon atom is sp^2 hybridized. In the second step, the carbocation immediately reacts with the nucleophile, a water molecule in this case, producing the conjugate acid of an alcohol (a protonated alcohol). The protonated alcohol dissociates immediately in the presence of excess water, forming the neutral alcohol and a hydronium ion.

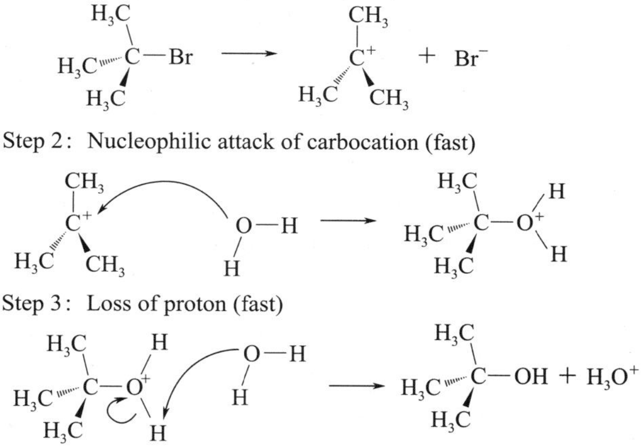

Figure 3.2 The S_N1 mechanism for the reaction between 2-bromo-2-methylpropane and water

This mechanism is designated S_N1. Again, the S indicates substitution and the N nucleophilic; in this case, the 1 indicates that the rate-determining step is unimolecular.

Apart from the rate law, what other evidence is there for the planar carbocation? Again, we can use a chiral haloalkane and investigate the chirality of the product. When one of the enantiomers of 3-bromo-3-methylhexane reacts with water, the product is a racemic mixture of the enantiomers of 3-methyl-3-hexanol. The carbocation intermediate formed in the rate-determining step is planar. The water molecular (nucleophile) can form a new bond on either face of the carbocation intermediate. This results in a mixture of the product enantiomers (Figure 3.3). The formation of a racemic mixture provides confirmation of the unimolecular rete-determining step in the S_N1 mechanism.

When is a mechanism S_N1 and when is it S_N2? To answer this question, we need to briefly discuss some kinetic studies carried out by Christopher Ingold and Edward Hughes in 1937. Measurement of rates of reaction under different sets of experimental conditions can provide insight into reaction mechanisms. For example, consider the rate of the following second-order reaction between a bromoalkene and the nucleophile Cl^-:

$$R-Br + Cl^- \xrightarrow{S_N 2} R-Cl + Br^-$$

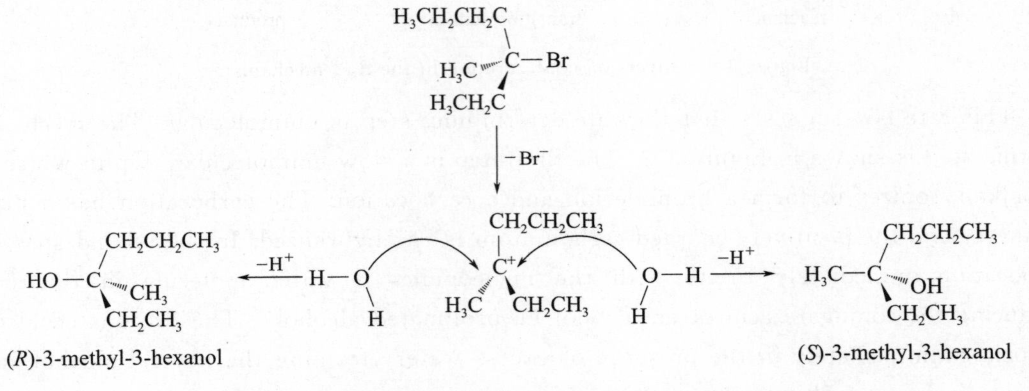

Figure 3.3 Formation of a racemic mixture in an $S_N 1$ reaction

From a thermodynamic point of view, the substitution of Br^- by Cl^- is favorable because Br^- is a weaker base than the Cl^- ion. The rate of reaction shows a dramatic dependence on the degree of branching of the α carbon. The order of the rate for this $S_N 2$ reaction: bromomethane > bromoethane > bromopropane > 2-bromopropane > 1-bromo-1-methylpropane.

In the $S_N 2$ mechanism, the nucleophile attacks the electrophilic center on the side opposite the leaving group, often called **backside attack**. Because the nucleophile attacks the backside of the α carbon, bulky substituents bonded to the α carbon will make it harder for the nucleophile to get the backside. The obstruction of the nucleophile from interacting with an electrophilic carbon atom is an example of steric hindrance. Backside attack of bromomethane is fastest because there is very little steric hindrance. However, the backside of 1,1-dimethylbromoethane is almost completely blocked from nucleophilic attack, and thus this compound does not undergo an $S_N 2$ reaction; nucleophilic substitution occurs by an $S_N 1$ mechanism instead.

What effect does alkyl substitution have on the rate of an $S_N 1$ reaction? When we look at a series of bromoalkanes reacting with water, a very different order of reactivity is observed:

$$R-Br + H_2O \longrightarrow R-OH + HBr$$

$S_N 1$ reactivity: $CH_3Br < CH_3CH_2Br < (CH_3)_2CHBr < (CH_3)_3CBr$

α carbon:　　methyl　　primary　　secondary　　tertiary

The order of reactivity, which is the opposite of that observed for the $S_N 2$ reaction, follows the order of carbocation stability. The reactivity is related to the stability of the carbocation because the rate-determining step in the $S_N 1$ reaction is the formation of the carbocation. The order of carbocation stability is as follows: $CH_3^+ < CH_3CH_2^+ < (CH_3)_2CH^+ < (CH_3)_3C^+$.

Thus, reagents or reaction conditions that favor the formation of a carbocation of a carbocation will increase the rate of the $S_N 1$ reaction.

It is important to emphasize, however, that carbocations are not particular stable, at least not in the usual sense. Carbocations are reactive intermediates with rather short lifetimes. A tertiary carbocation, such as $(CH_3)_3C^+$, has a lifetime of about 10^{-10} s in water. What is the explanation for the relative stabilities of methyl, primary, secondary, and tertiary carbocations? Alkyl groups stabilize carbocations and, in what role, they appear to be electron donating. However, the explanation of how alkyl groups stabilize a carbocation is a matter of some debate. One explanation is based on the concept of **hyperconjugation**, which is illustrated in Figure 3.4 for $CH_3CH_2^+$, a primary carbocation. The positive carbon is sp^2 hybridized. Around the positively charged carbon atom is a trigonal planar arrangement of atoms and an empty $2p$ orbital perpendicular to the plane of hybridization. The adjacent carbon atom is sp^3 hybridized and forms a σ bond with a hydrogen atom. The C—H bond can donate electron density to the empty $2p$ orbital, as suggested by the dashed arrow in Figure 3.4. If more alkyl groups are present in a carbocation, then more interactions of this type occur, and a greater degree of stabilization is the result. Thus, tertiary carbocations are more stable than secondary carbocations. Because no alkyl groups are present in a methyl carbocation, methyl carbocations are least stable; in fact, it is assumed they are never formed.

Figure 3.4　Stabilization of a carbocation through hyperconjugation

Other explanations for the stabilization of carbocations by alkyl groups are also used. However, all explanations have a common theme: the donation of electron density from filled orbitals that are aligned (or partially aligned) with the empty $2p$ orbital on the positively charged carbon atom of the carbocation. Irrespective of the explanation, alkyl groups help to stabilize a carbocation; thus, as the number of alkyl groups bonded to the α carbon increases, the stability of the carbocation also increases, as does the speed of the S_N1 reaction.

New Words and Expressions

carbocation /ˌkɑːbə(ʊ)ˈkatʌɪən/ n. 碳正离子
chiral /ˈtʃɪrəl/ adj. 手性的
electrophile /ɪˈlektrəfaɪl/ n. 亲电子试剂
enantiomer /ɪˈnæntiəmər/ n. 对映异构体
hybridize /ˈhaɪbrɪdaɪz/ v. 杂化
hyperconjugation /haɪpəkɒndʒʊˈɡeɪʃən/ n. 超共轭效应

leaving group n. 离去基团
mechanism /ˈmekənɪzəm/ n. 反应机理
nucleophile /ˈnjuːklɪəʊˌfaɪl/ n. 亲核试剂
nucleophilic substitution reaction n. 亲核取代反应
orbital /ˈɔːbɪtl/ n. 轨道
transition state n. 过渡状态

Notes

1. -phile 表示"爱好",如 nucleophile(亲核试剂),electrophile(亲电子试剂)。

2. This type of chemical reaction is called a nucleophilic substitution reaction because the hydroxide ions is a nucleophile—a reactant that seeks sites of low electron density in a molecule.

参考译文:因为氢氧根离子是亲核试剂——一种在分子中寻找低电子密度位置的反应物,所以这类化学反应被称为亲核取代反应。

3. In a nucleophilic substitution reaction, the bond between carbon and the leaving group is broken and a new bond is formed by using a lone pair from the nucleophile.

参考译文:在亲核取代反应过程中,碳原子与离去基团之间的键断开,而亲核试剂利用一对孤电子与这个碳原子同时生成一个新的键。

Questions

1. What are the nucleophilic substitution reactions? What are their reaction mechanisms?
2. What are nucleophiles and electrophiles?
3. How can we distinguish the reactions undergoing S_N1 or S_N2 reactions?

3.6 Isomerism

Isomers are different compounds that have the same chemical formula. Constitutional isomerism and stereoisomerism are two different types of isomerism.

3.6.1 Constitutional Isomerism

Constitutional isomers, also called **structural isomers**, occur when the same atoms can be connected in two or more different ways. For example, there are three different ways to arrange the atoms in a compound with the chemical formula C_5H_{12}. Constitutional isomers have distinct names and generally have different physical and chemical properties. Table 3.6 lists the three constitutional isomers of C_5H_{12} along with their boiling points for comparison.

Table 3.6 Boiling points of constitutional isomers of C_5H_{12}

Name	n-Pentane	Methylbutane (Isopentane)	2,2-Dimethylpropane (Neopentane)
Bond-line Structure			
Boiling point/℃	36.1	27.8	9.5

3.6.2 Stereoisomerism

Stereoisomers are *those that contain identical bonds but differ in the orientation of those bonds in space*. There are two types of stereoisomers: geometrical isomers and optical isomers. Geometrical isomers occur in compounds that have restricted rotation around a bond. For example, compounds that contain carbon-carbon double bonds can form geometrical isomers. Individual geometrical isomers have the same names but are distinguished by a prefix such as *cis-* or *trans-*. Dichloroethylene, ethylene in which two of the H atom (one on each C atom) have been replaced by Cl atoms, exists as two geometrical isomers. The isomer in which the Cl atoms both lie on the same side (above or below, in this example) of the double bond is called the **cis** isomer. The isomer in which the Cl atoms lie on opposite sides of the double bond is called the **trans** isomer (The compound in which both Cl atoms are attached to the same C atom is a constitutional isomer rather than a stereoisomer).

cis-dichloroethylene   *trans*-dichloroethylene   1,1-dichloroethylene

Geometrical isomers usually have different physical and chemical properties. *Trans* isomers tend to be more stable and are generally easier to synthesize than their *cis* counterparts. The existence of *cis* isomers in living system is a testament to how much better nature is at chemical synthesis than we are. Geometrical isomerism is sometimes of tremendous biological significance.

Stereoisomers that are mirror images of each other, but are not superimposable, are called **optical isomers.** Consider the hypothetical organic molecule CAEJG, which consists of a sp^3-hybridized carbon atom that is bonded to four different groups (A, E, J and G). It's mirror image appears identical to it, just as your right and left hands appear identical to each other. But, if you have ever tried to put a right-handed glove on your left hand, or vice versa, you know that your hands are not identical. Imagine rotating the molecule on the right so that it's A group and J group coincide with those of the molecule on the left. Doing so result in the E group of one molecule coinciding with the G group on the other. These two molecules are mirror images of each other, but they are not identical.

Molecules with nonsuperimposable mirror images are called **chiral**; and a pair of such mirror-image molecules are called **enantiomers**. Most of the chemical properties of enantiomers and all their physical properties (with the exception of the direction of rotation of plane-polarized light) are identical. Their chemical properties differ only in reactions that involve another chiral species, such as a chiral molecule or a receptor site that is shaped to fit only one enantiomer. Most biochemical processes consist of a series of chemically specific reactions that use chiral receptor sites to facilitate reaction by allowing only the specific reactants to fit and thus react.

In organic chemistry, it often is necessary to represent tetrahedral molecules (three-dimensional objects) on paper (a two-dimensional surface). By convention, this is done using solid lines to represent bonds that lie in the plane of the page, dashes to represent bonds that point behind the page, and wedge to represent bonds that point in front of the page.

One property of chiral molecules is that the two enantiomers rotate the plane of plane-polarized light in opposite directions; that is, they are optically active. Unlike ordinary light, which oscillates in all directions, plane-polarized light oscillates only in a single plane. We use a polarimeter to measure the rotation of polarized light by optical isomers. A beam of unpolarized light first passes through a Polaroid sheet, called the **polarizer**, and then through a sample tube containing a solution of an optically active, chiral species. As the polarized light passes through the sample tube, its plane of polarization is rotated either to the right or to the left. The amount of rotation can be measured by tuning the analyzer in the appropriate direction until minimal light transmission is achieved.

Minimal transmission occurs when the plane of polarization of the light is perpendicular to that of the analyzer through which it is viewed. This effect can be demonstrated using two pairs of polarized sunglasses. If the plane of polarization is rotated to the right, the isomer is said to be **dextrorotatory** and is label d; if it is rotated to the left, the isomer is called **levorotatory** and label l. Enantiomers always rotate the light by the same amount, but in opposite directions. Thus, in an equimolar mixture of both enantiomers, called a **racemic mixture**, the net rotation is zero.

A fascinating use of plane-polarized light is in 3-D movies. We see in three dimensions because our eyes view the world from slightly different positions. Our brains synthesize a three-dimensional (3-D) picture based on the two different pictures sent to it by our eyes. Modern 3-D movies make use of this phenomenon to make it seem as though objects on the screen are actually moving toward the viewer.

Three dimensional movies are filmed using two different cameras at slightly different angles to the action. Thus, there are actually two movies that must be shown to us simultaneously. To make sure that our two eyes receive two different perspectives, each movie is projected through a polarizer, which polarizes the two projections in directions perpendicular to each other.

Numerous processes important to biological function involve one enantiomer of a chiral compound. Many drugs, including thalidomide, are chiral with only one enantiomer having the desired properties. Some such drugs have been manufactured and marketed as racemic mixtures. It has become common, though, for drug companies to invest in **chiral switching**, the preparation of single-isomer versions of drugs originally marketed as racemic mixtures, in an effect to improve on existing therapies and to combat the revenue losses caused by generic drugs. A fairly high-profile of this is the single-isomer drug Nexium, the so-called purple pill. AstraZeneca, makers of Nexium, held a patent for the drug Prilosec, originally a prescription heartburn medication that is now available over the counter. Prilosec is a racemic mixture of the chiral compound omeprazole. Prior to 2002 expiration of its patent on Prilosec, AstraZeneca began producing and marketing Nexium, which contains only the therapeutically effective enantiomer (S)-omeprazole or esomeprazole.

Another example of chiral switching is that of the selective serotonin reuptake inhibitor (SSRI) antidepressant a Celexa, which was introduced to the market in 1998 by Forest Laboratories. Celexa is a racemic mixture of (R)-citalopram oxalate and (S)-citalopram oxalate. While only the (S) enantiomer has therapeutic antidepressant properties, both enantiomers contribute to the side effects of the drug and therefore limit effectiveness and patient tolerance. In 2002, the FDA approved Lexapro, a new antidepressant derived from Celexa but from which the therapeutically ineffective (R) enantiomer has been removed. The benefits of isolating the active isomer include smaller required dosages, reduced side effects, and a faster and better patient response to the drug.

Although the intellectual property laws regarding single-isomer drug patents are somewhat ambiguous, chiral switching has enabled some pharmaceutical companies to extend the time that they are able to market their popular prescriptions exclusively. Strictly speaking, the FDA does not consider a single enantiomer of an already approved chiral drug to be a "new chemical entity", which is a requirement for obtaining a patent on a compound. Early in the history of chiral switching, however, there was some disagreement among patent examiners regarding what constituted a new chemical entity, and patents were granted on single-isomer drugs that might not be granted today.

New Words and Expressions

antidepressant /ˌæntɪdɪˈpresnt/ n. 抗抑郁药
chiral switching n. 手性转换
citalopram oxalate n. 草酸西酞普兰（抗抑郁药）
constitutional isomerism n. 结构异构
dextrorotatory /ˌdekstrəʊˈrəʊtət(ə)rɪ/ adj. 右旋的
geometrical isomer n. 几何异构体
isomerism /aɪˈsɒmərɪzəm/ n. 同分异构
levorotatory /ˌliːvəʊˈrəʊtətərɪ/ adj. 左旋的

omeprazole /əʊˈmeprəzəʊl/ n. 奥美拉唑（胃药）
optical isomer n. 光学异构体
polarizer /ˈpəʊləraɪzə(r)/ n. 偏振光片
Polaroid sheet n. 偏振光片
racemic mixture n. 外消旋混合物
serotonin /ˌserəˈtəʊnɪn/ n. 血清素
stereoisomerism /ˌsterɪəʊaɪˈsɒmərɪzəm/ n. 立体异构
thalidomide /θəˈlɪdəmaɪd/ n. 沙利度胺（反应停）

Notes

1. To make sure that our two eyes receive two different perspectives, each movie is projected through a polarizer, which polarizes the two projections in directions perpendicular to each other.

参考译文：为了确保我们的两只眼睛接收到两种不同的透视效果，每部电影都通过偏振器进行投影，偏振器将产生两个偏振方向互相垂直的投影。

2. It has become common, though, for drug companies to invest in *chiral switching*, the preparation of single-isomer versions of drugs originally marketed as racemic mixtures, in an effort to improve on existing therapies and to combat the revenue losses caused by generic drugs.

参考译文：在成功制备药物的单一异构体前，药物都是以外消旋混合物的形式销售的。然而，对于制药公司来说，投资手性转换（技术）已成为一种普遍现象。其目的是改进现有的治疗方法，并"对抗"仿制药所造成的收入损失。

3. Although the intellectual property laws regarding single-isomer drug patents are somewhat ambiguous, chiral switching has enabled some pharmaceutical companies to extend the time that they are able to market their popular prescriptions exclusively.

参考译文：虽然关于单一异构体药物专利的知识产权法律（条文）有些含糊，但手性转换技术使一些制药公司能够延长他们独家销售其畅销处方的时间。

4. Early in the history of chiral switching, however, there was some disagreement among patent examiners regarding what constituted a new chemical entity, and patents were granted on single-isomer drugs that might not be granted today.

参考译文：然而，在手性转换的早期历史中，专利审查人员对于什么构成了一个新的化学个体存在一些分歧，并且对单异构体药物授予了专利，若以现今的技术水平来衡量，这些所谓单异构体的药物可能将不会授予任何专利。

Questions

1. What is the definition of isomers?
2. Give examples for constitutional isomers.
3. What are stereoisomers?
4. How can we know we get a single enantiomer instead of a racemic mixture?

3.7 Organic Polymers

Polymers are molecular compounds, either natural or synthetic, that are made up of many repeating units called **monomers**. The physical properties of these so-called macromolecules differ greatly from those of small, ordinary molecules.

The development of polymer chemistry began in the 1920s with the investigation of the puzzling behavior of some materials including rubber, wood, cotton, and gelatin. For example, when rubber, with the known empirical formula of C_5H_8, was dissolved in an organic solvent, the solution displayed several properties, including a higher than expected viscosity, which suggested that the dissolved compound had a very high molar mass. Despite the experimental evidence, though, scientists at the time were not ready to accept the idea that such giant molecules could exist. Instead, they postulated that materials such as rubber consisted of aggregates of small molecular units. Like C_5H_8 or $C_{10}H_{16}$, held together by intermolecular force. This misconception persisted for a number of years, until Hermann Staudinger clearly showed that these so-called aggregates, were, in fact, enormously large molecules, each of which contained many thousands of atoms held together by covalent bonds.

Once the structures of these macromolecules were understood, the way was open for the synthesis of polymers, which now pervade almost every aspect of our daily lives. The following discusses the reaction that result in polymer formation and some of the natural polymers that are important to biology.

3.7.1 Addition polymers

Addition polymers form when monomers such as ethylene join end to end to make polyethylene. Reactions of this type can be initiated by a radical—a species that contains an unpaired electron. The mechanism of addition polymerization is as follows:

(1) The radical, which is unstable because of its unpaired electron, attacks a carbon atom on an ethylene molecule. This is the **initiation** of the reaction.

(2) This attack would result in the carbon atom in question having more than eight electrons around it. To keep the carbon atom from having too many electrons around it, the double bond breaks.

(3) One of the electrons in the double bond, together with the electron from the radical, becomes a new bond between the ethylene molecule and the radical.

(4) The other electron remains with the other carbon atom, generating a new radical species.

(5) The new radical species, also unstable, attacks another ethylene molecule, causing the same sequence of events to happen again. The generation of a new, highly reactively radical species at each step is known as **propagation** of the reaction.

(6) Each step lengthens the chain of carbon atoms and results in the formation of a new radical, continuing the propagation of the reaction. Reactions such as this are known as chain reactions. They continue until the system runs out of ethylene molecules or unit the radical species encounters another radical species-resulting in **termination** of the reaction.

Initiation step:

$R\cdot + CH_2=CH_2 \longrightarrow R-CH_2-CH_2\cdot$

Propagation step:

$R-CH_2-CH_2\cdot + CH_2=CH_2 \longrightarrow R-CH_2-CH_2-CH_2-CH_2\cdot$

Termination step:

$R-[CH_2-CH_2]_n-CH_2-CH_2\cdot + \cdot R \longrightarrow R-[CH_2-CH_2]_n-CH_2-CH_2-R$

Some familiar and important additional polymers are listed in Table 3.7.

Table 3.7 Some additional polymers

Name	Monomer	Uses
Polyethylene	$CH_2=CH_2$	Plastic bags, bottle, toys
Polypropylene	$H_2C=CH-CH_3$	Carpeting, bottles
Polytetrafluoroethylene (Teflon®)	$CF_2=CF_2$	Nonstick coatings
Polyvinylchloride (PVC)	$H_2C=CH-Cl$	Water pipes, garden horses, plastic wrap
Polystyrene (Styrofoam)	$H_2C=CH-C_6H_5$	Packing material, insulation furniture

3.7.2 Condensation Polymers

Reaction in which two or more molecules become connected with the elimination of a small molecule, often water, are called **condensation reactions**. **Condensation polymers** form when molecules with two different functional groups combine, with the elimination of a small molecule, often water. Many condensation polymers are **copolymers**, meaning that they are made up of two or more different monomers.

$$RCOOH + HOR \longrightarrow RCOOR + H_2O$$

$$ROH + HOR \xrightarrow{H_2SO_4} ROR + H_2O$$

The first synthetic fiber, nylon 66, is a condensation copolymer of two molecules: one with carboxyl groups at each end (adipic acid) and one with amine groups at both ends (hexamethylenediamine).

$$H_2N-(CH_2)_6-NH_2 + HOOC-(CH_2)_4-COOH$$

$$\downarrow \text{Condensation}$$

$$H_2N-(CH_2)_6-\underset{H}{N}-\overset{O}{\underset{\|}{C}}-(CH_2)_4-COOH + H_2O$$

$$\downarrow \text{Further condensation reactions}$$

$$-(CH_2)_4-\overset{O}{\underset{\|}{C}}-\underset{H}{N}-(CH_2)_6-\underset{H}{N}-\overset{O}{\underset{\|}{C}}-(CH_2)_4-\overset{O}{\underset{\|}{C}}-\underset{H}{N}-(CH_2)_6-$$

Nylon was first made by Wallace Carothers at DuPont in 1931. The versatility of nylons is so great that the annual production of nylons and related substances now amounts to several billion pounds.

3.7.3 Control of Polymer Properties

One of the hallmarks of polymer chemistry is the ability to make useful materials. All plastic we use are formed from polymers. The ability to control the properties of materials comes from the ability to control their structures. We examine here means that influence properties unique to polymers: chain length, branching, and cross-linking.

3.7.3.1 *Chain Length*

The length of polymer chains influences properties such as a polymer's flexibility and its ability to dissolve in a solvent. In addition to polymerization, chain length is controlled by adjusting the ratio of initiator to monomer in the reaction mixture. When an addition polymerization reaction occurs, initiators start the reaction and monomers keep adding onto growing polymer chains until they are all reacted. If few initiator molecules are used, fewer chains will be formed, and each will add more monomers before the reaction stops. If many initiator molecules are used, more chains will form and the chains will have fewer monomer units each.

3.7.3.2 *Branching*

It is possible to use branching units in polymerization reactions that cause side chains to grow off the main polymer chain. This leads to bulky structures that do not pack closely together. Examples of this are high- and low-density polyethylene (HDPE and LDPE). HDPE is made up of straight-chain (not branched) polymer chains that pack closely together, resulting in a high-density solid. LDPE contains branched polyethylene, which cannot pack as tightly and results in a solid polymer with lower density and a more flexible structure.

3.7.3.3 *Crossing-Linking*

It is possible to link one chain to another using branches off one chain that bond to branches off an adjacent chain. These linkages are called cross-links. Materials containing cross-linked polymers are generally much tougher and harder than those consisting of non-cross-linked polymers. The toughness and hardness can be controlled by the amount of cross-linking.

3.7.4 Biological Polymers

Naturally occurring polymers include proteins, polysaccharide, and nucleic acids.

Proteins, polymers of amino acids, play an important role in nearly all biological processes. The human body contains an estimated 100,000 different kinds of proteins, each of which has a specific physiological function. An amino acid has both the carboxylic acid functional group and the amino functional group. Amino acids are joined together into chains when a condensation reaction occurs between a carboxyl group on one molecule and an amino group on another molecule.

$$H-{}^+N(H)(H)-C(H)(R_1)-C(=O)-O^- + H-{}^+N(H)(H)-C(H)(R_2)-C(=O)-O^- \rightleftharpoons H-{}^+N(H)(H)-C(H)(R_1)-C(=O)-N(H)-C(H)(R_2)-C(=O)-O^- + H_2O$$

The bonds that form between amino acid are called **peptide bonds**. Very long chains of amino acids assembled in this way are called **proteins**, while shorter chains are called **polypeptides**.

Amino acids consist of a central carbon atom bonded to four different groups: an amino group, a carboxyl group, a hydrogen atom, and an additional group consisting of carbon, hydrogen, and sometimes other elements such as nitrogen of sulfur. Proteins are made essentially from the 20 different amino acids. The identity of a protein depends on which of the 20 amino acids it contains and on the order in which the amino acids are assembled.

Polysaccharides are polymers of sugars such as glucose and fructose. Starch and cellulose are two polymers of glucose with slightly different linkages—and very different properties. In starch, glucose molecules are connected by what biochemists call α linkages. This enables animals, including humans, to digest the starch in such food as corn, wheat, potatoes, and rice.

starch

In cellulose, the glucose molecules are connected by β linkages. Digestion of cellulose requires enzymes that most animals do not have. Species that do digest cellulose, such as termites and ruminants (including cattle, sheep, and llamas), do so with the help of enzyme-producing symbiotic bacteria in the gut.

cellulose

Nucleic acid, which are polymers of nucleotides, play an important role in protein synthesis. There are two types of polymers of nucleic acids: **deoxyribonucleic acid (DNA)** and **ribonucleic acid (RNA)**. Each **nucleotide** in nucleic acid consists of a purine or pyrimidine base, a furanose sugar (deoxyribose for DNA; ribose for RNA), and a phosphate group. These molecules are among the largest known—they can have molar masses of up to tens of billions of grams. RNA molecules, on the other hand, typically have molar masses on the order of tens of thousands of grams. Despite their sizes, the composition of nucleic acids is relatively simple compared with proteins. Proteins consist of up to 20 different amino acids, whereas DNA and RNA consist of only four different nucleotides each.

New Words and Expressions

addition polymer n. 加成聚合物
amino acid /əˌmiːnəʊˈæsɪd/ n. 氨基酸
cellulose /ˈseljuləʊs/ n. 纤维素
condensation polymer n. 缩合聚合物
condensation reaction n. 缩合反应
deoxyribonucleic acid (DNA) n. 脱氧核糖核酸
fructose /ˈfrʌktəʊs/ n. 果糖
gelatin /ˈdʒelətɪn/ n. 明胶
glucose /ˈgluːkəʊs/ n. 葡萄糖
initiation /ɪˌnɪʃiˈeɪʃn/ n. 开始；启动
llamas /ˈlɑːmə/ n. 美洲驼
misconception /ˌmɪskənˈsepʃn/ n. 误解
monomer /ˈmɒnəmə(r)/ n. 单体
nucleic acid n. 核酸

nucleotide /ˈnjuːkliətaɪd/ n. 核苷酸
nylon /ˈnaɪlɒn/ n. 尼龙
peptide bond n. 肽键
pervade /pəˈveɪd/ v. 遍及
polymer /ˈpɒlɪmə(r)/ n. 聚合物
polysaccharide /ˌpɒliˈsækəraɪd/ n. 多糖
propagation /ˌprɒpəˈgeɪʃn/ n. 扩展
protein /ˈprəʊtiːn/ n. 蛋白质
puzzling /ˈpʌzlɪŋ/ adj. 令人迷惑的
ribonucleic acid (RNA) n. 核糖核酸
ruminant /ˈruːmɪnənt/ n. 反刍动物
starch /stɑːtʃ/ n. 淀粉
termination /ˌtɜːmɪˈneɪʃn/ n. 终止
termite /ˈtɜːmaɪt/ n. 白蚁

Notes

1. macro- 表示"巨大的，大量的"，如 macromolecule（大分子），macroscopic（宏观

的）。

2. co- 表示"共同，联合"，如 copolymer（共聚物），cooperation（合作），coworker（同事）。

3. de- 有多种意义，其中一种为"离开，出"，如 deoxyribonucleic acid（脱氧核糖核酸）。

4. Polymers are molecular compounds, either natural or synthetic, that are made up of many repeating units called monomers.

参考译文：无论是天然的还是合成的聚合物，它们都是由许多称为单体的重复单元组成的分子化合物。

5. For example, when rubber, with the known empirical formula of C_5H_8, was dissolved in an organic solvent, the solution displayed several properties, including a higher than expected viscosity, which suggested that the dissolved compound had a very high molar mass.

参考译文：例如，当已知实验式为 C_5H_8 的橡胶在有机溶剂中溶解时，该溶液显示出若干特性，包括高于预期的黏度，这表明溶解的化合物具有很高的分子量。

6. Amino acids are joined together into chains when a condensation reaction occurs between a carboxyl group on one molecule and an amino group on another molecule.

参考译文：当一个分子上的羧基和另一个分子上的氨基发生缩合反应时，氨基酸就连在一起形成链。

7. Amino acids consist of a central carbon atom bonded to four different groups: an amino group, a carboxyl group, a hydrogen atom, and an additional group consisting of carbon, hydrogen, and sometimes other elements such as nitrogen or sulfur.

参考译文：氨基酸由一个中心碳原子组成，它连着四个不同的基团：一个氨基基团、一个羧基基团、一个氢原子以及另外一个由碳、氢或者其他元素（如硫或氮）组成的基团。

Questions

1. Give the definition of a polymer.
2. How can a polymer be synthesized by a condensation reaction? Give an example.
3. Give examples of biopolymers.

Chapter 4

Analytical Chemistry

4.1 Introduction to Analytical Chemistry

There are two aspects that analytical chemistry is concerned with. The first is the chemical characterization of matter. The second is to answer two important questions: what is it (qualitative) and how much is it (quantitative). Chemicals make up everything we use or consume, and knowledge of the chemical composition of many substances is important in our daily lives. Analytical chemistry plays a very important role in nearly all aspects of chemistry, for instance, manufacturing, environmental, clinical, agricultural, forensic, metallurgical, and pharmaceutical chemistry. The nitrogen content of a fertilizer determines its value. Food must be examined for contaminants (e.g., pesticide residues) and for essential nutrients (e.g., vitamin content). The air in cities must be analyzed for toxic substances, such as carbon monoxide and sulfur dioxide. Most diseases are diagnosed by chemical analysis (for example, blood glucose is monitored in diabetics). The quality of manufactured products often depends on their proper chemical proportions, and measurement of the constituents is a necessary part of quality control. The carbon content of steel will determine its quality. The purity of drugs will determine their efficacy.

4.1.1 Qualitative Analysis

Qualitative analysis deals with the identification of elements, ions, or compounds present in a sample. In some cases, we may be interested in whether only a given substance is present. The sample may be in different states (solid, liquid, and gas) or in the form of a mixture. The presence of gunpowder residue on a hand generally requires only qualitative knowledge.

Qualitative tests may be performed by selective chemical reactions or with the use of instrumentation. For example, the formation of a white precipitate when adding a solution of silver nitrate to a dissolved sample indicates the presence of chloride ions. Colors produced in certain chemical reactions could be used to indicate the presence of classes of organic compounds, for example, ketones. Infrared spectra will give "fingerprints" of organic compounds or their functional groups.

There is a clear distinction between the terms **selective** and **specific**. A selective reaction or test is one that can occur with other substances but exhibits a degree of preference for the substance of interest. A specific reaction of test is one that occurs only with the substance of interest.

Unfortunately, few reactions are specific but many reactions exhibit selectively. Selectivity may be achieved by a number of strategies. Here are some examples:
- Sample preparation (e.g., extraction, precipitation).
- Target analyte derivatization (e.g., derivatize specific functional groups with detecting reagents).
- Instrumentation (selective detectors).
- Chromatography providing powerful separation.

4.1.2 Quantitative Analysis

One discipline of analytical chemistry deals with the determination of composition of materials—that is, the analysis of materials. The materials that one might analyze include food, water, air, hair, body fluids, pharmaceutical preparations, and so on. The analysis of materials is divided into qualitative and quantitative analysis. Identification of substances or species present in a material, such as lead(Ⅱ) ions in drinking water, uses qualitative analysis. On the other hand, the determination of the amount of a substance or species present in a material, for example, the amount of lead(Ⅱ) ions in one liter of drinking water, is involved in quantitative analysis.

4.1.2.1 *Gravimetric Analysis*

Gravimetric analysis is a type of quantitative analysis in which the amount of species in a material is determined by converting the species to a product that can be isolated completely and weighted. Precipitation reactions are frequently used in gravimetric analyses. In these reactions, you determine the amount of an ionic species by precipitating it from a solution. A precipitate, or solid formed in the reaction, is then filtered from the solution, dried, and weighted. The advantages of a gravimetric analysis are its simplicity (at least in theory) and its accuracy. The chief disadvantage is that it requires meticulous, time-consuming work. Because of this, whenever possible, chemists use modern instrumental methods.

As an example of a gravimetric analysis, consider the problem of determining the amount of lead in a sample of drinking water. Lead, if it occurs in the water, probably exists as the lead(Ⅱ) ion, Pb^{2+}. Lead(Ⅱ) sulfate is a very insoluble compound of lead(Ⅱ) ion.

When sodium sulfate, Na₂SO₄, is added to a solution containing Pb^{2+}, lead (Ⅱ) sulfate precipitates (that is, $PbSO_4$ comes out of the solution as a fine, crystalline solid). If you assume that the lead is present in solution as lead (Ⅱ) nitrate, you can write the following equation for the reaction:

$$Na_2SO_4(aq) + Pb(NO_3)_2(aq) \longrightarrow 2NaNO_3(aq) + PbSO_4(s)$$

You can separate the write precipitate of lead (Ⅱ) sulfate from the solution by filtration. Then you dry and weight the precipitate. After knowing the mass of the precipitate, you can calculate the amount of lead (Ⅱ) ions in a sample of drinking water.

4.1.2.2 *Volumetric analysis*

An important method for determining the amount of a particular substance is based on measuring the volume of reactant solution. Suppose substance X reacts in solution with substance Z. If you know the volume and concentration of a solution of Z that just reacts with substance X in a sample, you can determine the amount of X. **Titration** is a procedure for determining the amount of substance X by adding a carefully measured volume of a solution with known concentration of Z until the reaction of X and Z is just complete. **Volumetric analysis** is a method of analysis based on titration.

Let us consider a flask containing hydrochloric acid with an unknown amount of HCl being titrated with sodium hydroxide solution, NaOH, of known molarity. The reaction is

$$NaOH(aq) + HCl(aq) \longrightarrow NaCl(aq) + H_2O(l)$$

To the HCl, solution are added a few drops of phenolphthalein indicator. Phenolphthalein is colorless in the hydrochloric acid but turns pinks at the completion of the reaction of NaOH with HCl. Sodium hydroxide with a concentration of 0.207 mol/L is contained in a buret, a glass tube graduated to measure the volume of liquid delivered from the stopcock. The solution in the buret is added to the HCl solution in the flask until the phenolphthalein just changes from colorless to pink. At this point, the reaction is complete and the volume of NaOH that reacts with the HCl is read from the buret. This volume is then used to obtain the mass of HCl in the original solution.

New Words and Expressions

buret /bjʊˈret/ n. 滴定管
clinical /ˈklɪnɪkl/ adj. 临床的
diabetic /ˌdaɪəˈbetɪk/ n. 糖尿病患者
efficacy /ˈefɪkəsi/ n. 功效；效力
fingerprint /ˈfɪŋɡəprɪnt/ n. 指纹
forensic /fəˈrensɪk/ adj. 法医的
gravimetric analysis n. 重量分析
ketone /ˈkiːtəʊn/ n. 酮
metallurgical /ˌmetəˈlɜːdʒɪkl/ adj. 冶金的

meticulous /məˈtɪkjələs/ adj. 细心的
phenolphthalein /ˌfiːnɒlfˈθeɪlən/ n. 酚酞
qualitative /ˈkwɒlɪtətɪv/ adj. 定性的
quantitative /ˈkwɒntɪtətɪv/ adj. 定量的
residue /ˈrezɪdjuː/ n. 残留物；残渣
stopcock /ˈstɒpkɒk/ n. 旋塞；活栓
time-consuming /ˈtaɪm kənsjuːmɪŋ/ adj. 费时的
titration /taɪˈtreɪʃn/ n. 滴定
volumetric analysis n. 容量分析

Notes

1. 滴定管：buret（美式英语）与 burette（英式英语）。
2. 与滴定有关的词：titration（*n.* 滴定）；titrant（*n.* 滴定剂）；titrate（*vt.* 滴定）。
3. Gravimetric analysis is a type of quantitative analysis in which the amount of species in a material is determined by converting the species to a product that can be isolated completely and weighted.

参考译文：重量分析法是一种定量分析方法，其通过将物质转化为可完全分离和称重的产物来确定样品中各组分的含量。

4. Titration is a procedure for determining the amount of substance X by adding a carefully measured volume of a solution with known concentration of Z until the reaction of X and Z is just complete.

参考译文：滴定是一种确定物质 X 含量的方法，经由加入一个精确测量体积的已知浓度的溶液 Z，直至 X 和 Z 完全反应为止。

Questions

1. What are things which analytical chemistry is concerned with?
2. Give two examples for quantitative analysis.

4.2 Spectrochemical Methods

Spectrometry is one of the most widely used methods for analytical chemistry. Measurements for spectrometry can be made in some specific wavelength regions, such as infrared (0.78~300 μm), visible (380~780 nm), and ultraviolet (200~380 nm). The choice of wavelength region depends on some factors, such as availability of instruments, whether the analyte has a color or can be transformed into a colored derivative, whether it has functional groups which can absorb in the ultraviolet or infrared regions, and whether other absorbing species are present in solution. **Infrared (IR) spectrometry** is generally less suited for quantitative measurements but better suited for qualitative or **fingerprinting information** than are ultraviolet (UV) and visible spectrometry. Visible spectrometers are generally less expensive and are available than ultraviolet spectrometers.

4.2.1 Transitions between Internal Energy Levels

There are three basic processes by which a molecule can absorb radiation: **rotational transition**, **vibrational transition**, and **electronic transition**. In these processes, a molecule absorbs electromagnetic radiation, the energy of which equals the energy difference between a lower energy level and a higher energy level, depending on the types of transitions. The different energy levels are shown in Figure 4.1. The energy levels of rotation come from the molecule rotation about various axes. Atoms or groups of atoms within a molecule vibrate relative to each rotational can result in different vibration energy levels. In addition, there are different electronic energy levels. The energy difference between two adjacent energy level is in the order: electronic > vibrational > rotational. Rotational transitions occur at very low energies (in the energy range of microwave or far-infrared region), but vibrational transitions need higher energies in the near-infrared region, while electronic transitions require still higher energies (in the visible to ultraviolet region).

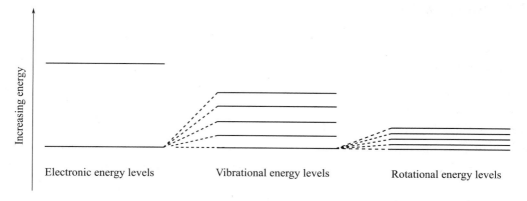

Figure 4.1 Energy level diagram

4.2.1.1 *Rotational Transitions*

Pure rotational transitions can take place in the far-infrared and microwave regions (*ca.* 100 μm to 10 cm), where the energy is insufficient to cause other transitions (vibrational or electronic transitions). The molecule at room temperature is usually populated in its lowest electronic state, called the **ground state**. Thus, the pure rotational transition will take place at the ground-state electronic level, although it is also possible be have an appreciable population of **excited states** of the molecule. When only rotational transitions occur, discrete absorption lines will be observed in the spectrum. Each line corresponds to a particular transition. Although fundamental information can be got about rotational energy levels of molecules, this region has less practical use.

4.2.1.2 *Vibrational Transitions*

In addition to rotational transitions, vibrational transitions take place in higher energies (smaller wavelengths) with different combinations of vibrational-rotational transitions. Each rotational level of the lowest vibrational level can be excited to different rotational levels of the excited vibrational level, each with several rotational levels. This leads to numerous discrete transitions, and results in a spectrum of peaks or unresolved fine structure. The peaks at different wavelengths relate to vibrational modes within the molecules.

4.2.1.3 *Electronic Transitions*

At still higher energies in the ultraviolet and visible (UV-vis) region, different levels of electronic transitions occur, and rotational and vibrational transitions are superimposed on these. This produces an even larger number of possible transitions. Although all the transitions take place in quantized steps corresponding to discrete wavelength, these individual wavelengths are too numerous and too close to resolve into the individual lines or vibrational peaks. Thus, a spectrum with broad bands of absorbed wavelengths is obtained generally.

4.2.2 Infrared Spectra

Infrared spectroscopy is very useful to obtain qualitative information about molecules. But the molecules must possess certain properties in order to undergo absorption.

Not all molecules can absorb in the infrared region. For absorption to occur, there must be a change in the **dipole moment (polarity)** of the molecule. A diatomic molecule must have a permanent dipole (polar covalent bond in which a pair of electrons is shared unequally) in order to absorb, but larger molecules do not. For instance, oxygen, $O=O$, cannot exhibit a diploe and will not absorb in the infrared region. On the other hand, an unsymmetrical diatomic molecule such as carbon monoxide does not have a permanent dipole and hence will absorb. Carbon dioxide, $O=C=O$, does not have a permanent diploe, but by vibration it may exhibit a dipole moment. Thus, in the vibration mode $O \Rightarrow C \Leftarrow O$, there is symmetry and no dipole moment. But in the mode $O \Leftarrow C \Leftarrow O$, there is a dipole moment and the molecule can absorb infrared radiation via an induced dipole.

Absorbing (vibrating) groups in the infrared region absorb within a certain wavelength region, and the exact wavelength will be influenced by neighboring groups. The absorption peaks are much sharper than in the ultraviolet or visible region and easier to identify. In addition, each molecule will have a complete absorption spectrum unique to that molecule, and so a "fingerprint" of the molecule is obtained. Catalogs of infrared spectra are available for a large number of compounds for comparison purposes. Mixtures of absorbing compounds will exhibit the combined spectra of compounds. Even so, it is often possible to identify the individual compounds from the absorption peaks of specific groups on the molecules. Typical functional groups that can be identified include alcohol hydroxyl, ester carbonyl, olefin, and aromatic unsaturated hydrocarbon groups. Table 4.1 summarizes regions where certain types of groups absorb. Absorption in the 6 to 15 μm region is very dependent on the molecular environment, and this is called the **fingerprint region**. A molecule can be identified by a comparison of its absorption in this region with cataloged known spectra.

Although the most important use of infrared spectroscopy is in identification and structure analysis, it is useful for quantitative analysis of complex mixtures of similar compounds because some absorption peaks for each compound will occur at a definite and selective wavelength, with intensities proportional to the concentration of absorbing species.

Table 4.1 Regions of the infrared spectrum

Wavenumber range / cm^{-1}	Types of Absorptions
3400~2800	O—H, N—H, C—H stretching
2250~2100	C≡N, C≡C stretching
1850~1600	C=O, C=N, C=C stretching
1600~1000	C—C, C—O, C—N stretching; various bending absorptions
1000~600	C—H bending

4.2.3 Electronic Absorption Spectra

The electronic transitions that occur in the visible and ultraviolet regions of the spectrum are due to the absorption of radiation by specific types of groups, bonds, and functional groups within the molecule. The wavelength of absorption and the intensity are dependent on the type. The wavelength of absorption is a measure of the energy required for the transition. Its intensity is dependent on the probability of the transition occurring when the electronic system and the radiation interact and on the polarity of the excited state.

4.2.3.1 *Kinds of Transitions*

Electrons in a molecule can be classified into four different types. (1) Closed-shell electrons that are not involved in bonding. These have very high excitation energies and do not contribute to absorption in the visible or UV regions. (2) Covalent single-bond electrons (σ, or sigma, electrons). These also possess too high an excitation energy to contribute to absorption of visible or UV radiation (e.g., single-valence bonds in saturated hydrocarbons, —CH_2—CH_2—). (3) Paired nonbonding outer-shell electrons (n electrons), such as those on N, O, S and halogens. These are less tightly held than σ electrons and can be excited by visible or UV radiation. (4) Electrons in π (pi) orbitals, for example, in double or triple bonds. These are the most readily excited and are responsible for a majority of electronic spectra in the visible and UV regions.

Electrons reside in orbitals. A molecule also possesses normally unoccupied orbitals called **antibonding orbitals**; these correspond to excited-state energy levels and are either σ^* or π^* orbitals. Hence, absorption of radiation results in an electronic transition to an antibonding orbital. The most common transitions are from π or n orbitals to antibonding π^* orbitals, and these are represented by $\pi \rightarrow \pi^*$ and $n \rightarrow \pi^*$ transitions, indicating a transition to an excited π^* state. The nonbonding n electron can also be promoted, at very short wavelengths, to an antibonding σ^* state: $n \rightarrow \sigma^*$. These occur at wavelengths less than 200 nm.

Examples of $\pi \rightarrow \pi^*$ and $n \rightarrow \pi^*$ transitions occur in ketones [$RC(O)R'$]. For example, acetone exhibits a high-intensity $\pi \rightarrow \pi^*$ transition and a low-intensity $n \rightarrow \pi^*$ transition in its absorption spectrum. An example of $n \rightarrow \pi^*$ transition occurs in ethers (R—O—R′). Since this occurs below 200 nm, ethers as well as thioether (R—S—R′), disulfides (R—S—S—R), alkyl amines (R—NH_2), and alkyl halides are transparent in the visible and UV regions; that is, they have no absorption bands in these regions.

The probability of π→π* transitions is greater than for n→π* transitions, and so the intensities of the absorption bands are greater for the former. Molar absorptivities, ε, at the band maximum for π→π* transition are typically 1000 to 100,000, while for n→π* transitions they are less than 1000; ε is direct measure of the intensities of the bands.

4.2.3.2 *Isolated Chromophores*

The absorbing groups in a molecule are called **chromophores**. A molecule containing a chromophore is called a **chromogen**. An **auxochrome** does not itself absorb radiation, but, if present in a molecule, it can enhance the absorption by a chromophore or shift the wavelength of absorption when attached to the chromophore. Examples are hydroxy groups, amino groups, and halogens. These possess unshared (n) electrons that can interact with the π electrons in the chromophore (n-π conjugation).

Spectral changes can be classed as follows: (1) **bathochromic shift**—absorption maximum shifted to longer wavelength; (2) **hypsochromic shift**—absorption maximum shifted to shorter wavelength; (3) **hyperchromism**—an increase in molar absorptivity; (4) **hypochromism**—a decrease in molar absorptivity.

In principle, the spectrum due to a chromophore is not markedly affected by minor structural changes elsewhere in the molecule. For example, acetone [$CH_3C(O)CH_3$] and 2-butanone [$CH_3C(O)CH_2CH_3$] give spectra similar in shape and intensity. If the alteration is major or is very close to the chromophore, then changes can be expected.

Similarly, the spectral effects of two isolated chromophores in a molecule (separated by at least two single bonds) are, in principle, independent and are additive. Hence, in the molecule CH_3CH_2CNS, an absorption maximum due to the CNS group occurs at 245 nm with an ε of 800. In the molecule $SNCCH_2CH_2CH_2CNS$, an absorption maximum occurs at 247 nm with approximately double the intensity (ε = 2000). Interaction between chromophores may perturb the electronic energy levels and alter the spectrum. Table 4.2 lists some common chromophores and their approximate wavelengths of maximum absorption.

Table 4.2 Electronic Absorption Bands for Representative Chromophores

Chromophore	System	$\lambda_{max}(\varepsilon_{max})$
Aldehyde	—CHO	270~285 (18~30)
Amine	—NH_2	195 (2800)
Azo	—N=N—	220~230 (1000~2000), 300~400 (10)
Ethylene	—C=C—	190 (8000)
Ketone	R_2—C(=O)—R_1	195 (100)
Nitrile	—ONO	210 (strong)
Nitro	—NO_2	210 (strong), 280~300 (11~18)
Anthracene		252 (199000), 375 (7900)
Benzene		184 (46700), 202 (6900), 255 (170)
Naphthalene		220 (11200), 275 (5600), 312 (175)

It should be noted that exact wavelengths of an absorption band and the probability of absorption (intensity) cannot be calculated, and the analyst always runs standards under carefully specified conditions (temperature, solvent, concentration, instrument type, etc). Modern instruments may have databases of standard spectra, and standard catalogs of spectra are available for reference.

4.2.3.3 *Conjugated Chromophores*

When double or triple bonds are separated by just one single bond each, they are said to be conjugated. The π orbitals overlap, which decreases the energy between adjacent orbitals. The result is bathochromic shift in the absorption spectrum and generally an increase in the intensity. The greater the degree of conjugation (i.e., several alternating double, or triple, and single bonds), the greater the shift. Conjugation of multiple bonds with nonbonding electrons (n-π conjugation) also results in spectral changes, for example, $Me_2C=CH-NO_2$.

Aromatic systems (containing phenyl or benzene groups) exhibit conjugation. The spectra are somewhat different, however, than in other conjugated systems, being more complex. Benzene absorbs strongly at 200 nm ($\varepsilon_{max}=6900$) with a weaker band at 230 to 270 nm ($\varepsilon_{max}=170$). The weaker band exhibits considerable fine structures, each peak being due to the influence of vibrational sublevels on the electronic transitions.

As substituted groups are added to the benzene ring, a smoothing of the fine structure generally results, with a bathochromic shift and increase in intensity. Hydroxy (—OH), methoxy (—OCH$_3$), amino (—NH$_2$), nitro (—NO$_2$), and aldehydic (—CHO) groups, for example, increase the absorption about 10-fold; this large effect is due to n-π conjugation. Halogens and methyl (—CH$_3$) groups act as auxochromes.

Polynuclear aromatic compounds (fused benzene rings), for example, naphthalene, have increased conjugation and so absorb at longer wavelengths. Naphthacene, having four rings, exhibits an absorption maximum at 575 nm and is blue.

In polyphenyl compounds, such as Ph—$(C_6H_4)_n$—Ph, *para*-linked molecules (1, 6 positions) are capable of resonance interactions (conjugation) over the entire system, and increase number of *para*-linked rings result in bathochromic shifts (e.g., from 250 to 320 nm in going from $n=0$ to $n=4$). In meta-linked molecules (1,3 positions), however, such conjugation is not possible and no appreciable shift occurs up to $n=16$. The intensity of absorption increases, however, due to the additive effects of the identical chromophores.

Many heterocyclic aromatic compounds, for example, pyridine, absorb in the UV region, and added substituents will cause spectral change, as for phenyl compounds.

Indicator dyes used for acid-base titration and redox titrations are extensively conjugated systems and therefore absorb in the visible region. Loss or addition of a proton or an electron will markedly change the electron distribution and hence the color.

4.2.3.4 *Metal Complexes*

The absorption of ultraviolet or visible radiation by a metal complex can be ascribed to one or more of the following transitions: (1) excitation of the metal ion, (2) excitation of the ligand, or (3) charge transfer transition. Excitation of the metal ion in a complex has a very low molar absorptivity (ε), on the order of 1 to 100, and is not useful for quantitative analysis. Most ligands used are organic chelating agents that exhibit the absorption properties discussed above, that is, can undergo π→π* and n→π* transitions. Complexation with a metal ion is similar to protonation of the molecule and will result in a change in the wavelength and intensity of absorption. These changes are slight in most cases.

The intense color of metal chelates is frequently due to charge transfer transitions. This is simply the movement of electrons from the metal ion to the ligand, or vice versa. Such transitions include promotion of electrons from π levels in the ligand or from σ bonding orbitals to the unoccupied orbitals of the metal ion, or promotion of σ-bonded electrons to unoccupied π orbitals of the ligand.

When such transition occurs, a redox reaction actually occurs between the metal ion and the ligand. Usually, the metal ion is reduced and the ligand is oxidized, and the wavelength (energy) of maximum absorption is related to the ease with which the exchange occurs. A metal ion in a lower oxidation state, complexed with a high electron affinity ligand, may be oxidized without destroying the complex. An important example is the 1,10-phenanthroline chelate of iron(Ⅱ).

Charge transfer transitions are extremely intense, with ε values typically 10000 to 100000; they occur in either the visible or UV regions. The intensity (ease of charge transfer) is increased by increasing the extent of conjugation in the ligand. Metal complexes of this type are intensely colored due to their high absorption and are suited for the detection and measurement of trade concentration of metals.

New Words and Expressions

auxochrome /ˈɔːksəkrəʊm/ n. 助色团
bathochromic shift n. 红移
chromogen /ˈkrəʊmədʒən/ n. 发色体；色原体
chromophore /ˈkrəʊməfɔː/ n. 生色团；发色团
dipole moment n. 偶极矩
electronic transition n. 电子跃迁
excited state n. 激发态
fingerprint region n. 指纹区
ground state n. 基态
heterocyclic /ˌhetərəʊˈsaɪklɪk/ adj. 杂环的

hypochromism /ˌhaɪpəkˈrəʊmɪzəm/ n. 减色
hypsochromic shift n. 蓝移
infrared spectrometry n. 红外光谱法
naphthacene /ˈnæfθəsiːn/ n. 稠四苯；并四苯
1,10-phenanthroline /fəˈnænθrəliːn/ n. 邻二氮杂菲（1,10-菲咯啉）
polyphenyl /ˌpɒliˈfiːnɪl/ adj. 聚苯的
pyridine /ˈpɪrɪdiːn/ n. 吡啶
rotational transition n. 转动跃迁
spectrometry /spekˈtrɒmɪtri/ n. 光谱法
vibrational transition n. 振动跃迁

Notes

1. hyper- 表示"上，超过"，如 hyperchromism（增色），hypertension（高血压）。

2. hypo- 表示"下,少,次",如 hypochromism(减色), hypothesis(假说), hypochlorous acid(次氯酸)。

3. -metry 表示"度量",如 spectrometry(光谱法), gravimetry(重量分析法), volumetry(容量分析法), voltammetry(伏安法), stoichiometry(化学计量)。

4. chromo- 表示"色",如 chromogen(发色体), chromophore(发色团) chromosome(染色体), chromatography(色谱法)。

5. Infrared (IR) spectrometry is generally less suited for quantitative measurements but better suited for qualitative or fingerprinting information than are ultraviolet (UV) and visible spectrometry.

参考译文:红外光谱法通常不太适合用于样品的定量测量,但比紫外光谱法和可见光谱法更适合于定性测量或获取指纹信息。

6. Although all the transitions take place in quantized steps corresponding to discrete wavelength, these individual wavelengths are too numerous and too close to resolve into the individual lines or vibrational peaks.

参考译文:虽然所有的跃迁都有相对应的波长,但这些个别的波长数量庞大且过于接近,以致无法分辨出单独的光谱线或振动峰。

7. Although the most important use of infrared spectroscopy is in identification and structure analysis, it is useful for quantitative analysis of complex mixtures of similar compounds because some absorption peaks for each compound will occur at a definite and selective wavelength, with intensities proportional to the concentration of absorbing species.

参考译文:虽然红外光谱最重要的用途是鉴定和结构分析,但它对于含有类似化合物的复杂混合物进行定量分析也是很有用的,因为每种化合物所产生的一些吸收峰会选择性地出现在一个特定的波长上,而且吸收强度与该化合物的浓度成正比。

8. The electronic transitions that occur in the visible and ultraviolet regions of the spectrum are due to the absorption of radiation by specific types of groups, bonds, and functional groups within the molecule.

参考译文:在光谱的可见和紫外区域发生的电子跃迁是由于分子中特定类型的基团、键和官能团对该区域辐射的吸收。

Questions

1. What are the spectrochemical methods for compounds?
2. Give examples for rotational transitions, vibrational transitions, and electronic transitions.

4.3 Chromatography

In general, methods for chemical analysis are at best selective; few, if any, are truly specific. Consequently, the separation of the analyte from potential interferences is more often than not a vital step in analytical procedures. Without question, the most widely used means of performing analytical separations is **chromatography**, a method that finds application to all branches of science. Column chromatography was invented and named by the Russian botanist Mikhail Tswett. He employed the technique to separate various plant pigments such as chlorophylls and xanthophylls by passing solutions of these compounds through a glass column packed with finely divided calcium carbonate. The separated species appeared as colored bands on the column, which accounts for the name he chose for the method (Greek *chroma* meaning "color" and *graphein* meaning "to write").

The applications of chromatography have grown explosively in the last seven decades, owing not only to the development of several new types of chromatographic techniques but also to the growing need by scientists for better methods for characterizing complex mixtures. The tremendous impact of these methods on science is attested by the 1952 Nobel Prize in Chemistry that was awarded to A. J. P. Martin and R. L. M. Synge for their discoveries in the field. Perhaps more impressive is a list of twelve Nobel Prize awards between 1937 and 1972 that were based upon work in which chromatography played a vital role. By now, this list has undoubtedly increased by a considerable number.

Chromatography encompassed a diverse and important group of methods that permit the scientist to separate closely related components of complex mixtures; many of these separations are impossible by other means. In all chromatographic separations the sample is dissolved in a **mobile phase**, which may be a gas, a liquid, or a supercritical fluid. This mobile phase is then forced through an immiscible **stationary phase**, which is fixed in place in a column or on a solid surface. The two phases are chosen so that the components of the sample distribute themselves between the mobile and stationary phase to varying degrees. Those components that are strongly retained by the stationary phase move only slowly with the flow of mobile phase. In contrast, components that are weakly held by the stationary phase travel rapidly. As a consequence of these differences in mobility, sample components separate into discrete bands that can be analyzed qualitatively and/or quantitatively.

Chromatographic processes can be classified according to the type of equilibration process involved, which is governed by the type of stationary phase. Various bases of equilibration are: (1) adsorption, (2) partition, (3) ion exchange, and (4) pore penetration.

4.3.1 Adsorption Chromatography

The stationary phase is a solid on which the sample components are adsorbed. The mobile phase may be a liquid (liquid-solid chromatography) or a gas (gas-solid chromatography); the components distribute between the two phases through a combination of sorption and desorption processes. Thin layer chromatography (TLC) is a special example of adsorption chromatography in which the stationary phase is a plane, in the form of a solid supported on an inert plate, and the mobile phase is a liquid.

4.3.2 Partition Chromatography

The stationary phase of partition chromatography is a liquid supported on an inert solid. Again, the mobile phase may be a liquid (liquid-liquid partition chromatography) or a gas (gas-liquid chromatography, GLC).

In a normal mode of operations of liquid-liquid partition, a polar stationary phase (e.g., methanol on silica) is used with a nonpolar mobile phase (e.g., hexane). This favors retention of polar compounds and elution of nonpolar compounds and is called **normal-phase chromatography**. If a nonpolar stationary phase is used, with a polar mobile phase, then nonpolar solutes are retained more and polar solutes more readily eluted. This is called **reversed-phase chromatography** and is actually the most widely used.

4.3.3 Ion Exchange and Size Exclusion Chromatography

Ion exchange chromatography uses an ion exchange resin as the stationary phase. The mechanism of separation is based on ion exchange equilibria. In size exclusion chromatography, solvated molecules are separated according to their size by their ability to penetrate a sieve-like structure (the stationary phase).

These are arbitrary classifications of chromatographic techniques, and some types of chromatography are considered together as a separate technique, such as gas chromatography for gas-solid and gas-liquid chromatography. In every case, successive equilibria are at work that determine to what extent the analyte stays behind or moves along with the eluent (mobile phase). In column chromatography, the column may be packed with small particles that act as the stationary phase (adsorption chromatography) or are coated with a thin layer or liquid phase (partition chromatography). In gas chromatography, the more common form today is a capillary column in which microparticles or a liquid are coated on the wall of the capillary tube.

4.3.4 Gas Chromatography

Gas chromatography (GC) is one of the most versatile and ubiquitous analytical techniques in the laboratory. It is widely used for the determination of organic compounds. The separation of benzene and cyclohexane (b.p. 80.1℃ and 80.8℃, respectively) is extremely simple by gas chromatography, but it is virtually impossible by conventional distillation. Although Martin and Synge invented liquid-liquid chromatography in 1941, the introduction of gas-liquid partition chromatography by James and Martin a decade later had a more immediate and large impact for two reasons. First, as opposed to manually operated liquid-liquid column chromatography, GC required instrumentation for application, which was developed by collaboration among chemists, engineers, and physicists; and analyses were much more rapid and done on a small scale. Second, at the time of its development, the petroleum industry was required to have improved analytical monitoring and immediately adopted GC. Within a few short years, GC was used for the analysis of almost every type of organic compounds.

Very complex mixtures can be separated by this technique. When coupled with mass spectrometry as a detection system, virtually positive identification of the eluted compounds is possible at very high sensitivity, creating a very powerful analytical system.

There are two types of GC: **gas-liquid (adsorption) chromatography** and **gas-liquid (partition) chromatography**. The more important of the two is gas-liquid chromatography (GLC) used in the form of capillary column.

4.3.5 Liquid Chromatography

Gas chromatography (GC), because of its speed and sensitivity and quite broad applicability, has been more widely used since its development than the various modes of liquid chromatography. But the latter techniques have potentially broader use because approximately 85% known compounds are not sufficiently volatile or stable to be separated by gas chromatography. The wealth of chromatographic theory accumulated, primarily from gas chromatography, has led to the development of techniques of **high-performance liquid chromatography (HPLC)** that rival gas chromatography in performance and allows separations and measurements to be made in a matter of minutes. The driving force for the rapid acceptance of gas chromatography in the 1950s was its immediate application to the petrochemical industry. Conversely, the applicability of HPLC to the pharmaceutical industry made it the mainstay of pharmaceutical laboratories in the 1970s, and today the HPLC market for new instruments is larger than more mature GC market.

New Words and Expressions

attest /əˈtest/ v. 证实
capillary tube n. 毛细管
chlorophyll /ˈklɒrəfɪl/ n. 叶绿素
chromatography /ˌkrəʊməˈtɒgrəfi/ n. 色谱法
column chromatography n. 柱色谱
gas chromatography n. 气相色谱法
interference /ˌɪntəˈfɪərəns/ n. 干扰
liquid chromatography n. 液相色谱法

mainstay /ˈmeɪnsteɪ/ n. 支柱
mobile phase n. 流动相
normal-phase chromatography n. 正相色谱法
reversed-phase chromatography n. 反相色谱法
stationary phase n. 固定相
ubiquitous /juːˈbɪkwɪtəs/ adj. 十分普遍的
versatile /ˈvɜːsətaɪl/ adj. 多用途的；多功能的
xanthophyll /ˈzænθəˌfɪl/ n. 叶黄素

Notes

1. The applications of chromatography have grown explosively in the last seven decades,

owing not only to the development of several new types of chromatographic techniques but also to the growing need by scientists for better methods for characterizing complex mixtures.

参考译文：在过去的70年里，色谱技术的应用飞速发展，这不仅是由于几种新型色谱技术的发展，还有科学家们对更好地表征复杂混合物方法的日益增长的需求。

2. When coupled with mass spectrometry as a detection system, virtually positive identification of the eluted compounds is possible at very high sensitivity, creating a very powerful analytical system.

参考译文：当结合质谱作为检测系统时，洗脱下的化合物几乎都可以在非常高的灵敏度下进行鉴定，因而它们创建了一个非常强大的分析系统。

3. The wealth of chromatographic theory accumulated, primarily from gas chromatography, has led to the development of techniques of high-performance liquid chromatography (HPLC) that rival gas chromatography in performance and allows separations and measurements to be made in a matter of minutes.

参考译文：多年来积累丰富的色谱理论知识（主要来自气相色谱）引领了高效液相色谱（HPLC）技术的发展，其性能可与气相色谱相媲美，并使得分离和测量可以在几分钟内完成。

Questions

1. Classify the different chromatographic techniques, and give examples of principal types of applications.

2. What are the differences between normal-phase and reversed-phase chromatography?

4.4 Thin-Layer Chromatography

Thin-layer chromatography (TLC) is a planar form of chromatography useful for wide-scale qualitative analysis screening and can also be used for quantitative analysis. The stationary phase is a thin layer of finely divided adsorbent supported on a glass or aluminum plate, or plastic strip. Any of the solids used in column liquid chromatography can be used, provided a suitable binder can be found for good adherence to the plate.

A sample is spotted onto the plate with a micropipet, and the chromatogram is developed by placing the bottom of the plate or strip (but not the sample spot) in a suitable solvent. The solvent is drawn up the plate by capillary action, and the sample components move up the plate at different rates, depending on their solubility and their degree of retention by the stationary phase. Following development, the individual solute spots are noted or are made visible by treatment with a reagent that forms a colored derivative. The spots will generally move at a certain fraction of the rate at which the solvent moves, and they are characterized by the R_f value: $R_f =$ distance solute moves/distance solvent front moves, where the distances are measured from the center of where the sample was spotted at the bottom of the plate. The solvent front will be a line across the plate. The distance of the solute moves is measured at the center of the solute spot or at its maximum intensity, if tailing occurs. The value R_f is characteristic for a given stationary phase and solvent combination. Because of slight variation in plates, it is always a good idea to determine the R_f value on each set of plates.

4.4.1 Developing the Chromatogram

A thin layer pencil line is drawn across the plate a few centimeters from the bottom, and the sample is spotted on this for future reference in R_f measurements. The spot must be made as small as possible for maximum separation and minimum tailing. It is best done dropwise with a warm air blower (e.g., a hair dryer) to evaporate the solvent after each drop. The plate is placed in a chamber with its end dipping in the developing solvent. A closed (presaturated) chamber must be used to saturate the atmosphere with the solvent and prevent it from evaporating from the surface of the plate as it moves up. The developing may take 10 min to 1 h., but it requires no operation time. The amount of development will depend on the complexity of the mixture of solutes being separated. If a wide plate is used, several samples and standards can be spotted along the bottom and developed simultaneously.

Development times of 5 min can be accomplished by using small microscope slides for TLC plates, and preliminary separation with these can be conveniently used to determine the optimum developing conditions. Typically, sample spots should be 2~5 mm in diameter.

A principal advantage of this technique is that greater separating power can be achieved using two-dimensional thin-layer chromatography: a large square TLC plate is used, and the sample is spotted at a bottom corner of the plate. After development with a given solvent system, the plate is turn 90° and further development is obtained with a second solvent system. Thus, if two or more solutes are not completely resolved with the first solvent, it may be possible to resolve them with a second solvent. Proper control of the pH is often important in achieving efficient separations.

4.4.2 Detection of the Spots

If the solutes fluoresce (e.g., aromatic compounds), they can be detected by shining an ultraviolet light on the plane. A pencil line is drawn around the spots for permanent identification. Color-developing reagents are often used. For example, amino acids and amines are detected by spraying the plate with a solution of ninhydrin, which is converted to blue or purple. After the spots are identified, they may be scraped off and the solutes washed off (eluted) and determined quantitatively by a micromethod.

Frequently, colorless or nonfluorescent spots can be visualized by exposing the developed plate to iodine vapor. The iodine vapor interacts with the sample components, either chemically or by solubility, to produce a color. Thin-layer plates and sheets are commercially available that incorporate a fluorescent dye in the powdered adsorbent. When held under ultraviolet light, dark spots appear where sample spots occur due to quenching of the plate fluorescence.

A common technique for organic compounds is spraying the plate with a sulfuric acid solution and then heating it to char the compounds and develop black spots. This precludes quantitative analysis by scraping the spots off the plate and eluting for measurement.

4.4.3 Stationary Phases for TLC

The stationary phase consists of a finely divided powder (particle size $10 \sim 50$ μm). It can be an adsorbent, an ion exchanger, or a molecular sieve, or it can serve as the support for a liquid film. An aqueous slurry of the powder is prepared, usually with a binder such as plaster of paris, gypsum, or poly (vinyl alcohol) to help it adhere to the backing material. The slurry is spread on the plate in a thin film, typically $0.1 \sim 0.3$ mm, using a spreading adapter to assure uniform thickness. Adapters are commercially available. The solvent is evaporated off and adsorbents are activated by placing in an oven at 110 ℃ for several hours. Commercially prepared plates and strips on plastic are available.

The most commonly used stationary phases are **adsorbents**. Silica gel, alumina, and powdered cellulose are the most popular. Silica gel particles contain hydroxyl groups on their surface which will hydrogen bond with polar molecules. Adsorbed water prevents other polar molecules from reaching the surface, so the gel is activated by heating to remove the adsorbed water. Alumina also contains hydroxyl groups or oxygen atoms. Alumina is preferred for polar compounds such as amino acids and sugars. Magnesium silicate, calcium silicate, and activated charcoal may also be used as adsorbents. Adsorbents are sometimes not activated by heating, in which case the residual water acts as the stationary phase.

Thin-film **liquid stationary phases** can be prepared for separation by liquid-liquid partition chromatography. The film, commonly water, is supported on materials such as silica gel on diatomaceous earth, as in column chromatography. Either silica gel or diatomaceous earth may be silanized to convert the surfaces to nonpolar methyl groups for reversed-phase thin-layer chromatography.

Ion exchange resins are available in particle sizes of $40 \sim 80$ μm, suitable for preparing thin-layer plates. Examples are Dowex 50W strong-acid cation exchange and Dowex 1 strong-base anion exchange resins, usually in the sodium or hydrogen or the chloride forms, respectively. An aqueous slurry of six parts resin to one part cellulose powder is suitable for spreading into a $0.2 \sim 0.3$mm layer. The gel is soaked in water for about 3 days to complete the swelling, and then spread on the plate. The plates are not dried, but stored wet. The capillary action through these molecular sieves is much slower than with most other thin layers, typically only $1 \sim 2$ cm/h, and so development takes $8 \sim 10$ h, compared to about 30 min for other stationary phases.

4.4.4 Mobile Phase for TLC

In adsorption chromatography, the eluting power of solvents increases in the order of their polarities (e.g., from hexane to acetone to alcohol to water). A single solvent, or at most two or three solvents, should be used whenever possible, because mixed solvents tend to chromatograph as they move up the thin layer, causing a continual change in the solvent composition with distance on the plate. This may result in varying R_f values depending on how far the spots are allowed to move up the plane.

The developing solvent must be of high purity. The presence of small amounts of water or other impurities can produce irreproducible chromatograms.

4.4.5 Quantitative Measurements

The powerful resolving power of two-dimensional thin-layer chromatography has been combined with quantitative measurements by optically measuring the density of chromatographic spots. This can be done by measuring the transmittance of light through the chromatographic plate or the reflectance of light, which is attenuated by the analyte color. Or, fluorescence intensity may be measured upon illumination with ultraviolet radiation. Full spectrum recording and multiple wavelength scanning (with diode arrays) capabilities are commercially available.

4.4.6 High-Performance Thin-Layer Chromatography (HPTLC)

The power of thin-layer chromatography has been enhanced by consideration of chromatography principles to improve the speed and efficiency of separation and by the development of instrumentation to automate sample application, development of the chromatogram, and detection, including accurate and precise *in situ* quantitation as mentioned above. The use of a very fine particle layer results in faster and more efficient separations. The particle size has a narrower distribution range, with an average size of 5 μm, instead of the average 20 μm for conventional TLC. Mechanical applications permit reproducible application and reduction in the diameter of the starting spots. Smaller volume sample are used compared with conventional TLC, about one-tenth, and separation times are reduced by a factor of 10. In addition to precoated silica gel layers, a range of chemically bonded phases, similar to those used in normal and reversed phase high-performance liquid chromatography, are available.

The very fine particles used in HPTLC slows the movement of the mobile phase after a relatively short distance. To overcome this limitation, a "forced-flow" technique has been employed, using a pressurized chamber. The mobile phase is delivered with the aid of a pump at a constant velocity through a slit in a plastic sheet covering the stationary phase.

Modern thin-layer chromatography can be complementary to HPLC. It allows the processing of many samples in parallel, providing low-cost analysis of simple mixtures for which the sample workload is high. The TLC plates acts as "storage detectors" of the analyte if they are saved.

New Words and Expressions

activated charcoal n. 活性炭
adsorbent /əd'zɔ:bənt/ n. 吸附剂
alumina /ə'lu:mɪnə/ n. 矾土
calcium silicate n. 硅酸钙
capillary action n. 毛细管现象
cellulose /'seljuləʊs/ n. 纤维素
chamber /'tʃeɪmbə(r)/ n. 室
char /tʃɑ:(r)/ v. (使) 烧黑
chromatogram /'krəʊmətəˌɡræm/ n. 色谱图
diatomaceous earth n. 硅藻土

diode array n. 二极管阵列
fluoresce /flʊə'res/ n. 荧光
gypsum /'dʒɪpsəm/ n. 石膏
magnesium silicate n. 硅酸镁
ninhydrin /nɪn'haɪdrɪn/ n. 茚三酮
plaster of paris n. 熟石膏
silica gel n. 硅胶
slurry /'slʌri/ n. 泥浆
strip /strɪp/ n. 条；带
thin-layer chromatography n. 薄层色谱法

Notes

1. A sample is spotted onto the plate with a micropipet, and the chromatogram is developed by placing the bottom of the plate or strip (but not the sample spot) in a suitable solvent.

参考译文：用微量移液管将样品点在平板上，然后将平板或板条（而不是样品点）的底部置于合适的溶剂中展开，形成色谱图。

2. The solvent is drawn up the plate by capillary action, and the sample components move up the plate at different rates, depending on their solubility and their degree of retention by the stationary phase.

参考译文：溶剂通过毛细管作用被吸进板中，样品组分根据其溶解度及在固定相中的滞留程度，以不同的速度在平板中向上移动。

3. A principal advantage of this technique is that greater separating power can be achieved using two-dimensional thin-layer chromatography: a large square TLC plate is used, and the sample is spotted at a bottom corner of the plate.

参考译文：这种技术的一个主要优点是使用二维薄层色谱可以获得更大的分离能力，即使用一个大的方形薄层色谱板并且样品被点在板的底角。

4. A single solvent, or at most two or three solvents, should be used whenever possible, because mixed solvents tend to chromatograph as they move up the thin layer, causing a continual change in the solvent composition with distance on the plate.

参考译文：应尽可能使用单一溶剂或最多使用两种或三种溶剂，因为混合溶剂在薄层平板上移动时会分离，导致溶剂成分随板上距离产生持续变化的现象。

5. The power of thin-layer chromatography has been enhanced by consideration of

chromatography principles to improve the speed and efficiency of separation and by the development of instrumentation to automate sample application, development of the chromatogram, and detection, including accurate and precise *in situ* quantitation as mentioned above.

参考译文：薄层色谱分离能力可以经由使用色谱原理（提高分离的速度和效率）、开发自动化样品应用以及开发色谱图和检测（包括上述准确和精确的原位定量）来增强。

Questions

1. Describe the principle of TLC.
2. What are the differences between TLC and HPTLC?

4.5 NMR Spectroscopy

Infrared spectroscopy can be used to determine the functional groups present in a compound, and mass spectrometry provides the masses of a molecule and its coherent fragments. With rare exceptions, however, neither of these techniques give enough information to define a complete structure. Another form of spectroscopy, nuclear magnetic resonance (NMR) spectroscopy, enables the chemists to probe molecular structure in much greater detail. Using NMR, sometimes in conjunction with other forms of spectroscopy, but often by itself, a chemist can usually determine a complete molecular structure in a very short time. In the period since its introduction in the late 1950s, NMR spectroscopy has revolutionized not only organic chemistry but also other fields, such as biology and material science.

4.5.1 Principles

One of the most important uses of NMR spectroscopy is to examine the hydrogens in organic compounds. This type of NMR spectroscopy is based on the magnetic properties of the hydrogen nucleus that result from its **nuclear spin.**

Just as electrons have two allowed spin states, designated by the quantum numbers $+1/2$ and $-1/2$, some nuclei also have spin. The hydrogen nucleus ^1H, that is, the proton, has a nuclear spin that also can assume either of two values, designated by quantum numbers $+1/2$ and $-1/2$.

The physical significance of nuclear spin is that the nucleus acts like a tiny magnet. This means that hydrogen nuclei, whether alone or in a compound, respond to a magnetic field. When a compound containing hydrogens is placed in a magnet field, its hydrogen nuclei become magnetized.

To see what causes this magnetization, let us represent the hydrogen nuclei in a chemical sample with arrows indicating their magnetic ("north-south") polarity. In the absence of a magnetic field, the nuclear magnetic poles are oriented randomly. After a magnet field is applied, the magnetic poles of nuclei with spin of $+1/2$ are oriented parallel to the magnet field, and those of nuclei with spin of $-1/2$ are oriented antiparallel to the field.

The most important effect of the magnetic field for NMR is how it affects the energies of the two spin states. In the absence of a field, the two spin states have the same energies. But when a magnetic field is applied, the two spin states have different energies: the $+1/2$ spin state has lower energy than the $-1/2$ spin state. In addition, the populations of the two spin states are in rapid equilibrium. The spin equilibrium, like a chemical equilibrium, favors the state with lower energy. Thus, after the field is applied, more protons have spin $+1/2$ than spin $-1/2$. For every million protons in a sample, the state with spin $+1/2$ typically contains an excess of only ten to twenty protons—certainly not a large difference, but enough to be detected. Because the magnetic poles of the nuclei with spin $+1/2$ are aligned with the magnetic field, and because there are more of these nuclei, the sample has a small net magnetization in the direction of the field.

The energy difference ΔE between the two spin states for a given proton depends on the intensity of the magnetic field B_0 at the proton. The larger the field, the greater is ΔE.

Here is where things stand: Molecules of a sample are situated in a magnetic field; each nucleus is in one of two spin states that differ in energy by an amount ΔE; and more nuclei have spin $+1/2$. If the sample is now subjected to electromagnetic radiation of energy exactly equal to ΔE, this energy is absorbed by some of the nuclei in the $+1/2$ spin state. The absorbed energy causes these nuclei to invert or "flip" their spins and assume a more energetic state with spin $-1/2$.

This energy absorption by nuclei in a magnetic field is termed **nuclear magnetic resonance** and can be detected in a type of absorption spectrophotometer called a nuclear magnetic resonance spectrometer, or **NMR spectrometer**. The study of this absorption is called **NMR spectroscopy**.

To summarize: for nuclei to absorb energy, they must have a nuclear spin and must be situated in a magnetic field. Once these two conditions are met, then the nuclei can be examined by an absorption spectroscopy experiment. The absorption of energy corresponds physically to the "flipping" of nuclear spins from the $+1/2$ to the $-1/2$ spin state.

Even with the large magnetic fields conventionally used in NMR spectroscopy, the energy of the radiation required for the absorption experiment is very small—on the order of 0.02 J/mol or 0.005 cal/mol. The electromagnetic radiation with this energy lies in the FM area of the radiofrequency, or "RF", region of the electromagnetic spectrum. Consequently, FM radio waves are the radiation used in NMR spectroscopy. Typical values of the frequency used in NMR experiments are between 60×10^6 sec^{-1} and 500×10^6 sec^{-1}, or between 60 and 500 MHz Indeed, an ordinary FM radio receiver located near an NMR spectrometer and tuned to the appropriate frequency can produce audible sounds associated with an NMR experiment.

Any nucleus with a spin can be detected by NMR. Since proton have nuclear spin, NMR can be used to detect and study the hydrogen nuclei in any compound. ^{12}C, however, has no nuclear spin and therefore does not give an NMR signal. Other nuclei with spin are ^{19}F, ^{31}P, and the carbon isotope ^{13}C. These nuclei all have spins of $\pm 1/2$ and can be observed with NMR. Some nuclei also have spin values other than $\pm 1/2$ ^{14}N and deuterium (^2H), for example, can also be detected by NMR, but these nuclei have three allowed spin states of $\pm 1/2$ and 0. Because almost all organic compounds contain hydrogens, proton NMR spectroscopy is especially useful to the organic chemist. Despite the low abundance of ^{13}C, advances in instrumental techniques have made ^{13}C NMR a tool of major importance for the organic chemist.

4.5.2 The NMR Spectrum

In a typical NMR experiment, radiation with a frequency of 60 MHz is applied to the sample. This is called the operating frequency, or applied frequency, and is symbolized by v_0. Then the magnetic field of the instrument is applied and slowly increased over time. This increases the energy separation between the spin energy levels. Now, remember that energy and frequency are related by $E = hv$. When the energy separation between the spin energy levels exactly corresponds to a frequency of 60 MHz, energies absorbed from the applied frequency. This absorption is registered as a peak on the spectrum. Consequently, an NMR spectrum is a plot of energy absorption vs. magnet field strength. (Alternatively, the magnetic field could be fixed and the applied frequency varied to get the proper match.)

When a chemical compound is subjected to a high magnetic field in an NMR spectrometer, the field B_0 at a given proton within the compound is not the same as the field provided by the spectrometer, B_{ext}. The reason for this difference is that electrons circulating in the vicinity of a proton provide their own magnetic fields that oppose the external field. The external field B_{ext} has to be strong enough to overcome the fields of the circulating electrons in order to make B_0 larger enough for absorption to occur. (You might say that the external magnetic field has to "fight through" the field of the circulating electrons.) In addition, the field provided by the circulating electrons is different for protons in different chemical environments. This means, in turn, the magnetic field strength required for NMR absorption depends on the chemical environment of the proton. Turning this idea around, the strength of the field at which a proton absorbs energy gives information about its chemical environment.

Keeping the idea of the previous theory, let's examine an actual NMR spectrum. Absorption is plotted on the vertical axis in arbitrary units, increasing from bottom to top; that is, absorption peaks are registered as upward deflections in the spectrum. Absorption peaks are termed absorptions, lines, or resonances. Tetramethylsilane [TMS; $(CH_3)_4Si$] is added to the sample to provide an internal reference absorption from which the positions of all other absorptions are measured.

The strength of the applied field B_{ext} is plotted on the horizontal axis of an NMR spectrum, and increases to the right, or to the upfield direction. Absorption positions are always cited by their **chemical shifts**. If the absorption frequency for a proton of interest is v, and that of the standard TMS is v_{TMS} then the chemical shift of the proton is defined as the difference between its absorption position and that of TMS:

$$\text{chemical shift (in Hz)} = \nu - \nu_{TMS}$$

Notice that when $\nu = \nu_{TMS}$, the chemical shift is zero. That is, the chemical shift of TMS is set to zero by definition. The chemical shift scale is shown on the upper horizontal axis of the spectrum in units of hertz (Hz).

Although hertz is a unit of frequency, it can also be used as a unit of field strength because the energy difference between spin energy levels is proportional to the field strength. Remember that the energy of the radiation required to "flip" nuclear spins is equal to this energy difference, and that the energy and frequency of this radiation are proportional according to $E = h\nu$.

The chemical shift in Hz is proportional to the operating frequency.

$$\text{Chemical shift (in Hz)} = \nu - \nu_{TMS} = \delta \nu_0$$

In this equation, the frequencies ν and ν_{TMS} are in Hz and ν_0 is in MHz. Each proton has a characteristic δ value that is independent of the operating frequency. This equation says, for example, that if ν_0 is tripled, then the chemical shifts of all protons are also tripled. Thus, at an operating frequency of 180 MHz, the chemical shifts of both absorptions would be three times as great as they are at an operating frequency of 60 MHz.

The interpretation of NMR spectra often requires the use of tables of chemical shift data. Such table are most useful when chemical shifts are presented in a form that is independent of the operating frequency. The quantity δ can be expressed:

$$\delta = \frac{\nu - \nu_{TMS}}{\nu_0} = \frac{\text{chemical shift in Hz}}{\text{operating frequency in MHz}}$$

Because the units of the numerator are Hz and the units of the denominator are MHz, the units of δ are parts per millions, abbreviated ppm (1 ppm = 10^{-6}). Notice that the ppm scale increases in the downfield direction, which is toward lower magnet field. NMR spectra taken at different chemical operating frequencies have the same chemical shift scale in ppm but a different chemical shift scale in Hz.

In most organic molecules, protons give NMR absorptions over a chemical shift range of about 0~10 ppm downfield from TMS. Indeed, TMS was chosen as a standard not only because it is volatile and inert, but also because its NMR absorption occurs at a higher field than the absorptions of most common organic compounds. (The rare absorptions that occur at higher field than TMS are assigned negative δ values.)

New Words and Expressions

chemical shift n. 化学位移
NMR spectroscopy n. 核磁共振光谱
nuclear magnetic resonance n. 核磁共振
nuclear spin n. 核自旋

tetramethylsilane /ˈtetrəˌmeθilˈsilein/ n. 四甲硅烷
upfield /ˌʌpˈfiːld/ adj. 向高磁场的

Notes

1. Using NMR, sometimes in conjunction with other forms of spectroscopy, but often by itself, a chemist can usually determine a complete molecular structure in a very short time.

参考译文：通常单独利用核磁共振，有时结合其他光谱学，化学家就可以在很短的时间内确定一个完整的分子结构。

2. Absorption is plotted on the vertical axis in arbitrary units, increasing from bottom to top; that is, absorption peaks are registered as upward deflections in the spectrum.

参考译文：吸收强度以任意单位绘制在纵坐标轴上，从下往上递增；也就是说，光谱中的吸收峰为向上突出的峰。

Questions

1. What is the chemical-shift difference in ppm of two signals separated by 45 Hz at each of the following operating frequencies? (a) 60 MHz, (b) 300 MHz.

2. For ^{1}H NMR, what compounds is used as an internal standard?

4.6 Mass Spectrometry

In contrast to other spectroscopic techniques, mass spectrometry does not involve the absorption of electromagnetic radiation but operates on a completely different principle. Mass spectrometry is used to determine molecular masses, and it is the most important technique used for this purpose. It also has some use in determining molecular structure.

In a **mass spectrometer**, a compound is first vaporized in a vacuum and converted into ions, which then separated and detected. The most common ionization technique involves bombarding the compound with high-energy (70 electron-volts or 6700 kJ/mol) electrons. This technique is called **electron impact ionization** (EI). Since this energy is much greater than the bond energies of chemical bonds, some fairly drastic things happen when a molecule is subjected to such conditions. One thing that happens is that an electron is ejected from the molecule. For example, if methane is treated in this manner, it loses an electron from one of the C—H bonds.

$$\begin{array}{c} H \\ H:\ddot{C}:H \\ H \end{array} + e^- \longrightarrow \begin{array}{c} H \\ H:\overset{+}{\ddot{C}}:H \\ H \end{array} + 2e^-$$

The product of this reaction is sometimes abbreviated as follows:

$$\begin{array}{c} H \\ H:\ddot{C}:H \\ H \end{array} \text{ abbreviated as } CH_4^{\scriptsize{+}}$$

The symbol $\overset{+}{\cdot}$ means that the molecule is both a radical (a species with an unpaired electron) and a cation—a **radical cation**. The species $CH_4^{+\cdot}$ is called the methane radical cation.

Following its formulation, the methane radical cation decomposes in a series of reactions called **fragmentation reactions**. In one such reaction, it loses a hydrogen atom to generate the methyl cation, a carbocation.

$$CH_4^{+\cdot} \longrightarrow {}^+CH_3 + H\cdot$$
$$\text{mass=16} \quad \textbf{methyl cation}$$
$$\text{mass=15}$$

The hydrogen atom carries the unpaired electron, and the methyl cation carries the charge. The process can be represented with the free-radical (fishhook) arrow formation as follows:

$$H-\underset{H}{\overset{H}{C{:}H}} \longrightarrow H-\underset{H}{\overset{H}{C^+}} + H\cdot$$

Alternatively, the unpaired electron may remain associated with the carbon atom; in this case, the products of the fragmentation are a methyl radical and a proton.

$$CH_4^{+\cdot} \longrightarrow \cdot CH_3 \quad + H^+$$
$$\text{mass=16} \quad \textbf{methyl radical} \quad \text{mass=1}$$

Further decomposition reactions give fragments of progressively smaller mass. (Show how these occur by using the fishhook formalism.)

$$CH_3^+ \longrightarrow CH_2^+ + H\cdot$$
$$\text{mass=14}$$
$$CH_2^+ \longrightarrow CH^+ + H\cdot$$
$$\text{mass=13}$$
$$CH^+ \longrightarrow C^{+\cdot} + H\cdot$$
$$\text{mass=12}$$

Thus, methane undergoes fragmentation in the mass spectrometer to give several positively charged fragment ions of differing mass: $CH_4^{+\cdot}$, CH_3^+, $CH_2^{+\cdot}$, CH^+, $C^{+\cdot}$, and H^+. In the mass spectrometer, these fragment ions are separated according to their **mass-to-charge ratio**, m/z ($m=$ mass, $z=$ the charge of the fragment). Since most ions formed in the mass spectrometer have unit charge, the m/z value can generally be taken as the mass of the ion. A **mass spectrum** is *a graph of the relative amount of each ion (called the relative abundance) as a function of the ionic mass (or m/z)*. Note that only ions are detected by the mass spectrometer, and natural molecules and radicals do not appear as peaks in the mass spectrum. The mass spectrum of methane shows peaks at $m/z = 16, 15, 14, 13, 12$, and 1, corresponding to the various ionic species ($CH_4^{+\cdot}$, CH_3^+, $CH_2^{+\cdot}$, CH^+, $C^{+\cdot}$, and H^+) that are produced from methane by electron ejection and fragmentation.

The mass spectrum can be determined for any molecule that can be vaporized in a high vacuum, and this includes most organic compounds. We can use mass spectrum to determine the molecular mass of an unknown compound or to determine the structure (or partial structure) of an unknown compound by an analysis of the fragment ions in the spectrum.

The ion derived from electron ejection before any fragmentation takes place is known as the **molecular ion**, and abbreviated M. The molecular ion occurs at an m/z value equal to the molecular mass of the sample molecule. Thus, in the mass spectrum of methane, the molecular ion occurs at $m/z=16$. In the mass spectrum of decane, the molecular ion occurs at $m/z=142$. Except for peaks due to isotopes, the molecular ion peak is the peak of highest m/z in any ordinary mass spectrum.

The **base peak** is the ion of greatest relative abundance. The base peak is arbitrarily assigned a relative abundance of 100%, and the other peaks in the mass spectrum are scaled relative to it. In the mass spectrum of methane, the base peak is the same as the molecular ion, but in the mass spectrum of decane the base peak occurs at $m/z=43$. In the latter spectrum and most others, the molecular ion and the base peak are different.

The mass spectrum of methane shows a small but real peak at $m/z=17$, a mass one unit higher than the molecular mass. This peak is termed an $M+1$ peak, because it occurs one mass higher than the molecular ion (M). This ion occurs because chemically pure methane is readily a mixture of compounds containing the various isotope of carbon and hydrogen.

$$\text{methane} = {}^{12}CH_4, \quad {}^{13}CH_4, \quad {}^{12}CDH_3, \text{ etc}$$
$$m/z = \quad 16 \qquad 17 \qquad 17$$

The isotopes of several elements and their natural abundances are listed in Table 4.3. Possible sources of the $m/z=17$ peak for methane are $^{13}CH_4$ and $^{12}CDH_3$. Each isotopic compound contributes a peak with a relative abundance in proportion to its amount. In turn, the amount of each isotope compound is directly related to the natural abundance of the isotope involved. The relative abundance of a peak due to ^{12}C methane and the peak due to the presence of a ^{13}C isotope is then given by the following equation:

$$\text{Relative abundance} = \left(\frac{\text{abundance of } {}^{13}C \text{ peak}}{\text{abundance of } {}^{12}C \text{ peak}}\right)$$

$$= (\text{number of carbons}) \times \left(\frac{\text{natural abundance of } {}^{13}C}{\text{natural abundance of } {}^{12}C}\right)$$

$$= (\text{number of carbons}) \times \left(\frac{0.0110}{0.9890}\right)$$

$$= (\text{number of carbons}) \times 0.0111$$

Since methane has only one carbon, the $m/z=17$ ($M+1$) peak due to $^{13}CH_4$ is about 1.1% of the $m/z=16$, or M, peak. A similar calculation can be made for deuterium.

$$\text{Relative abundance} = (\text{number of hydrogens}) \times \left(\frac{\text{natural abundance of } {}^{2}H}{\text{natural abundance of } {}^{1}H}\right)$$

$$= 4 \times \left(\frac{0.00015}{0.99985}\right) = 0.0006$$

Table 4.3 Exact masses and isotopic abundance of some isotopes

Isotope	Exact mass (Abundance/%)
$^1H, {}^2H$	1.007825 (99.985), 2.0140 (1.015)
$^{12}C, {}^{13}C$	12.0000 (98.90), 13.00335 (1.10)
$^{16}O, {}^{17}O, {}^{18}O$	15.99491 (99.759), 16.99913 (0.037), 17.99916 (0.204)
$^{28}Si, {}^{29}Si, {}^{30}Si$	27.97693 (92.21), 28.97649 (4.67), 29.97377 (3.10)
$^{35}Cl, {}^{37}Cl$	34.96885 (75.77), 36.96590 (24.23)
$^{79}Br, {}^{81}Br$	78.91834 (50.69), 80.91629 (49.31)

Thus, the CDH$_3$ naturally present in methane contributes 0.06% to the isotopic peak. Because the contribution of deuterium is negligible, ^{13}C is the major isotope contribute to the $M+1$ peak.

In a compound containing more than one carbon, the $M+1$ peak is larger relative to the M peak because there is 1.1% probability that each carbon in the molecule will be present as ^{13}C. For example, cyclohexane has six carbons, and the abundance of its $M+1$ ion relative to that of its molecular ion should be $6×(1.1\%)=6.6\%$. In the mass spectrum of cyclohexane, the molecular ion has a relative abundance of about 70%; that of the $M+1$ ion is calculated to be $(0.066)×(70\%)=4.6\%$, which corresponds closely to the value observed. (With careful measurement, it is possible to use these isotope peaks to estimate the number of carbons in an unknown compound.) Not only the molecular ion peak, but also every other peak in the mass spectrum has isotopic peaks.

Several elements of importance in organic chemistry have isotopes with significant natural abundance. Table 4.3 shows that silicon has significant $M+1$ and $M+2$ contribution, and the halogens chlorine as well as bromine have very important $M+2$ contributions. In fact, the naturally occuring form of the element bromine consists of about equal amounts of ^{79}Br and ^{81}Br. The mixture of isotopes leaves a characteristic trail in the mass spectrum that can be used to diagnose the presence of the element.

Although isotopes such as ^{13}C and ^{18}O are normally present in small amounts in organic compounds, it is possible to synthesize compounds that are selectively enriched with these and other isotopes. Isotopes are espeically useful because they provide specific labels at particular atoms without changing their chemical properties. One use of such compounds is to determine the fate of specific atoms in deciding between mechanisms. Another use is to provide nonradiative isotopes for biological metabolic studies, which deal with the fates of chemical compounds when they react in biological systems. When a compound has been isotopically enriched, isotopic peaks are much larger than they are normally. Mass spectometry is used to measure quantitatively the amount of such isotopes present in labeled compounds.

New Words and Expressions

base peak n. 基峰
electron impact ionization n. 电子轰击离子化
fragmentation reaction n. 裂片反应
mass spectrometer n. 质谱仪

mass spectrometry n. 质谱法
mass-to-charge ratio n. 质荷比
molecular ion peak n. 分子离子峰
radical cation n. 自由基阳离子

Notes

1. In a mass spectrometer, a compound is first vaporized in a vacuum and converted

into ions, which then separated and detected.

参考译文：在质谱仪中，化合物首先在真空中蒸发并转化为离子，然后进行分离和检测。

2. A mass spectrum is a graph of the relative amount of each ion (called the relative abundance) as a function of the ionic mass (or m/z).

参考译文：质谱是每个离子的相对量（称为相对丰度）随离子质量（或 m/z）变化的分布图。

Questions

1. The mass spectrum of tetramethylsilane, $(CH_3)_4Si$, has a base peak at $m/z=73$. Calculate the relative abundance of the isotopic peaks at $m/z=74$ and 75.

2. Predict the appearance of the molecular ion peak(s) in the mass spectrum of chloromethane (Assume that the molecular ion is the base peak).

Chapter 5

Physical Chemistry

5.1 Introduction to Physical Chemistry

Physical chemistry is the study of the underlying physical principles that govern the properties and behavior of chemical systems.

A chemical system can be studied from either a point of view of microscopic or macroscopic. The **microscopic** viewpoint is based on the concept of molecules. The **macroscopic** viewpoint studies large-scale properties of matter without explicit use of the molecule concept.

We can divide physical chemistry into four areas: thermodynamics, quantum chemistry, statistical mechanics, and kinetics. **Thermodynamics** is a macroscopic science that studies the interrelationships of the various equilibrium properties of a system and the changes in equilibrium properties in processes.

Molecules and the particles (electrons and nuclei) that compose them do not obey classical mechanics. Instead, their motions are governed by the law of quantum mechanics. **Quantum chemistry** is the application of quantum mechanics to atomic structure, molecular bonding, and spectroscopy.

The macroscopic science of thermodynamics is a consequence of what is happening at a molecular (microscopic) level. The molecular and macroscopic levels are related to each other by the branch of science called **statistical mechanics**. Statistical mechanics gives insight into why the laws of thermodynamics hold and allows calculation of macroscopic thermodynamic properties from molecular properties.

Kinetics is the study of rate processes such as chemical reactions, diffusion, and the flow of change in an electrochemical cell. The theory of rate processes is not as well developed as the theories of thermodynamics, quantum mechanics, and statistical mechanics. Kinetics uses relevant portions of thermodynamics, quantum chemistry, and statistical mechanics.

The principles of physical chemistry provide a framework for all branches of chemistry.

Organic chemists use kinetics studies to figure out the mechanisms of reactions, use quantum-chemistry calculation to study the structures and stabilities of reaction intermediates, use symmetry rules deduced from quantum chemistry to predict the course of many reactions, and use nuclear-magnetic-resonance (NMR) and infrared spectroscopy to help determine the structure of compounds. Inorganic chemists use quantum chemistry and spectroscopy to study bonding. Analytical chemists use spectroscopy to analyze samples. Biochemists use kinetics to study rates of enzyme-catalyzed reactions; use thermodynamics to study biological energy transformations, osmosis, and membrane equilibrium, and to determine molecular weights of biological molecules; use spectroscopy to study processes at the molecular level (for example, intramolecular motions in proteins are studied using NMR); and use X-ray diffraction to determine the structures of proteins and nucleic acids.

Environmental chemists use thermodynamics to find the equilibrium composition of lakes and streams, use chemical kinetics to study the reactions of pollutants in the environment. Chemical engineers use thermodynamics to predict the equilibrium composition of reaction mixtures, use kinetics to calculate how fast product will be performed, and use principles of thermodynamics phase equilibria to design separation procedures such as fractional distillation. Geochemists use thermodynamic phase diagrams to understand process in the earth. Polymer chemists use thermodynamic, kinetics, and statistical mechanics to investigate the kinetics of polymerization, the molecular weights of polymers, the flow of polymer solutions, and the distribution of conformations of a polymer molecule.

Widespread recognition of physical chemistry as a discipline began in 1887 with the founding of the *Zeitschrift für Physikalische Chemie* by Wilherm Ostwald with J. H. van't Hoff as coeditor. Ostwald investigated chemical equilibrium, chemical kinetics, and solutions and wrote the first textbook of physical chemistry. He was instrumental in drawing attention to Gibbs' pioneering work in chemical thermodynamics and was the first to nominate Einstein for a Nobel Prize. Surprisingly, Ostwald argued against the atomic theory of matter and did not accept the reality of atoms and molecules until 1908. Ostwald, van't Hoff, and Arrhenius are generally regarded as the founders of physical chemistry.

In its early years, physical chemistry research was done mainly at the macroscopic level. With the discovery of the laws of quantum mechanics in 1925—1926, emphasis began to shift to the molecular level. Nowadays, the power of physical chemistry has been greatly increased by experimental techniques that study properties and processes at the molecular level and by fast computers that (a) process and analyze data of spectroscopy and X-ray crystallography experiments, (b) accurately calculate properties of molecules that are not too large, and (c) perform simulations of collections of hundreds of molecules.

The prefix *nano* is widely used in such terms as nanoscience, nanotechnology, nanomaterials, nanoscale, etc. A nanoscale (or nanoscopic) system is one with at least one dimension in the range 2~100 nm, where 1 nm = 10^{-9} m. (Atomic diameters are typically 0.1~0.3 nm.) A nanoscale system typically contains thousands of atoms. The intensive properties of a nanoscale system commonly depend on its size and differ substantially from those of a macroscopic system of the same composition. For example, macroscopic solid gold is yellow, is a good electrical conductor, melts at 1336 K, and is chemically unreactive; however, gold nanoparticles of radius 2.5 nm melt at 930 K, and catalyze many reactions; gold nanoparticles of 100 nm radius are purple-pink, of 20 nm radius are red, and of 1 nm radius are orange; gold particles of 1 nm or smaller radius are electrical insulations. The term mesoscopic is sometimes used to refer to systems larger than nanoscopic but smaller than macroscopic. Thus, we have the progressively larger size levels: atomic→nanoscopic→mesoscopic→macroscopic.

5.1.1 Thermodynamics

Thermodynamics (from the Greek words for "heat" and "power") is the study of heat, work, energy, and the changes they produce in the states of system. In a broader sense, thermodynamics studies the relationships between the macroscopic properties of as system. A key property in thermodynamics is temperature and thermodynamics is sometimes defined as the study of the relationship of temperature to the macroscopic properties of matter.

Equilibrium thermodynamics deals with systems in equilibrium. (**Irreversible thermodynamics** deal with nonequilibrium systems and rate processes.) Equilibrium thermodynamics is a macroscopic science and is independent of any theories of molecular structure. Strictly speaking, the word "molecule" is not part of the vocabulary of thermodynamics. However, we won't adopt a purist attitude but will often use molecular concepts to help us understand thermodynamics. Thermodynamics does not apply to systems that contain only a few molecules; a system must contain a great many molecules for it to be treated thermodynamically.

5.1.2 Kinetics

The study of rate processes is called **kinetics** or **dynamics**. Kinetics is one of the four branches of physical chemistry. A system may be out of equilibrium because matter or energy or both are being transported between the system and its surroundings or between one part of the system and another. Such processes are **transport processes**, and the branch of kinetics that studies the rates and mechanism of transports processes is **physical kinetics**. Even though neither matter nor energy is being transported through space, a system may be not of equilibrium because certain chemical species in the system are reacting to produce other species. The branch of kinetics that studies the rates and mechanisms of chemical reactions is **chemical kinetics** or **reaction kinetics**. A reacting system is not in equilibrium, so reaction kinetics is not part of thermodynamics but is a branch of kinetics.

5.1.3 Quantum Mechanics

Quantum chemistry applies quantum mechanics to chemistry. Unlike thermodynamics, quantum mechanics deals with system that are not part of everyday macroscopic experience, and the formulation of quantum mechanics is quite mathematical and abstract. This abstractness takes a while to get used to, and it is natural to feel somewhat uneasy when first learning quantum mechanics.

Essentially all of chemistry is a consequence of the laws of quantum mechanics. If we want to understand chemistry at the fundamental level of electrons, atoms, and molecules, we must understand quantum mechanics. Quantities such as the heat of combustion of butane, the 25℃ entropy of liquid water, the reaction rate of H_2 with O_2 gases at specified conditions, the equilibrium constants of chemical reactions, absorption spectra of coordination compounds, the NMR spectra of organic compounds, the nature of the products formed when organic compounds react, the shape a protein molecule folds into when it is formed in a cell, the structure and function of DNA are all a consequence of quantum mechanics.

Quantum mechanics was used to develop many concepts that helped explain chemical properties. However, because of the very difficult calculations needed to apply quantum mechanics to chemical systems, quantum mechanics was of a little practical value in accurately calculating the properties of chemical systems for many years after its discovery. Nowadays, however, the extraordinary computational power of modern computers allows quantum mechanical calculations to give accurate chemical predictions in many systems of real chemical interest. As computers become even more powerful and applications of quantum mechanics in chemistry increase, the need for all chemists to be familiar with quantum mechanics will increase.

5.1.4 Statistical Mechanics

The link between quantum mechanics and thermodynamics is provided by **statistical mechanics**, whose aim is to deduce the macroscopic properties of matter from the properties of the molecules composing the system. Typical macroscopic properties are chemical reaction rate, dielectric constant, electrical conductivity, entropy, heat capacity, internal energy, surface tension, and viscosity. Molecular properties include molecular masses, molecular geometries, intramolecular forces (which determine the molecular vibration frequencies), and intermolecular forces. Because of the huge number of molecules in a macroscopic system, one use statistical methods instead of attempting to consider the motion of each molecule in the system.

Statistical mechanics originated in the work of Maxwell and Boltzmann on the kinetic theory of gases (1860—1900). Major advances in the theory and the methods of calculation were made by Gibbs in his 1902 book *Elementary Principles in Statistical Mechanics* and by Einstein in a series of papers (1902—1904). Since quantum mechanics had not yet been established, these workers assumed that the system's molecules obeyed classical mechanics. This led to incorrect results in some cases. For example, calculated C_v's of gases of polyatomic molecules disagreed with experiments. When quantum mechanics was discovered, the necessary modifications in statistical mechanics were easily made.

New Words and Expressions

dynamics /daɪˈnæmɪks/ n. 动力学
equilibrium /ˌiːkwɪˈlɪbriəm/ n. 平衡
explicit /ɪkˈsplɪsɪt/ adj. 清楚明白的
instrumental /ˌɪnstrəˈmentl/ adj. 起重要作用的
interrelationship /ˌɪntərɪˈleɪʃənʃɪp/ n. 相互关联
irreversible /ˌɪrɪˈvɜːsəbl/ adj. 不可逆的

kinetics /kɪˈnetɪks/ n. 动力学
macroscopic /ˌmækrəʊˈskɒpɪk/ adj. 宏观的
microscopic /ˌmaɪkrəˈskɒpɪk/ adj. 微观的
quantum chemistry n. 量子化学
statistical mechanics n. 统计力学
thermodynamics /ˌθɜːməʊdaɪˈnæmɪks/ n. 热力学

Notes

1. 以-ics 结尾的学科：thermodynamics（热力学），mechanics（力学），kinetics（动力学），physics（物理学），mathematics（数学），economics（经济学）。

2. equilibria（平衡）为 equilibrium 的复数形式。其他以 um 结尾的单数名词，其复数以 a 结尾的例子有：quantum/quanta（量子），datum/data（数据），medium/media（媒体，媒介），bacterium/bacteria（细菌）。

3. Thermodynamics is a macroscopic science that studies the interrelationships of the various equilibrium properties of a system and the changes in equilibrium properties in processes.

参考译文：热力学是一门宏观科学，它研究系统内各种平衡性质之间的相互关系和过程中平衡性质的变化规律。

4. Statistical mechanics gives insight into why the laws of thermodynamics hold and allows calculation of macroscopic thermodynamic properties from molecular properties.

参考译文：统计力学让我们深入了解为什么热力学定律能够成立，并可以从分子性质计算宏观热力学性质。

5. The link between quantum mechanics and thermodynamics is provided by statistical mechanics, whose aim is to deduce the macroscopic properties of matter from the properties of the molecules composing the system.

参考译文：统计力学将量子力学和热力学之间联系起来，其目的是从组成系统的分子的性质推断物质的宏观性质。

Questions

What areas can be divided in physical chemistry?

5.2 Chemical Equilibrium

Chemical reactions often seem to stop before they are complete. Such reactions are reversible. That is, the original reactant form products, but then the products react with themselves to give back the original reactants. Actually, two reactions are occurring, and the eventual result is a mixture of reactants and products, rather than simply a mixture of products.

Consider the gaseous reaction in which carbon monoxide and hydrogen react to produce methane and steam

$$CO(g) + 3H_2(g) \longrightarrow CH_4(g) + H_2O(g)$$

This reaction, which requires a catalyst to occur at a reasonable rate, is called **catalytic methanation**. It is a potentially useful reaction because it can be used to produce methane from coal or organic wastes from which carbon monoxide can be obtained. Methane, in turn, can be used as a fuel and as a starting material for the production of organic chemicals. Although, at the moment, natural gas (which is mostly methane) is plentiful, analysts predict that oil and natural gas production will peak sometime this century. Then the demand for other sources of fuels and organic chemicals, such as coal and organic wastes, will rise.

Catalytic methanation is a reversible reaction, and depending on the reaction conditions, the final reaction mixture will have varying amounts of the products methane and steam, as well as the starting substances carbon monoxide and hydrogen. It is also possible to start with methane and steam and, under the right conditions, form a mixture that is predominantly carbon monoxide and hydrogen. The process is called **steam reforming**.

$$CH_4(g) + H_2O(g) \longrightarrow CO(g) + 3H_2(g)$$

The product mixture of CO and H_2 (synthesis gas) is used to prepare a number of industrial chemicals.

The process of catalytic methanation and steam reforming illustrate the reversibility of chemical reactions. An important question is, "what conditions favor the production of CH_4 and H_2O from CO and H_2, and what conditions favor the production of CO and H_2 from CH_4 and H_2O?"

Certain reactions (such as catalytic methanation) appear to stop before they are complete. The reaction mixture ceases to change in any of its properties and consists of both reactants and products in definite concentrations. Such a reaction mixture is said to have reached **chemical equilibrium**. Other types of equilibria include equilibrium between a liquid and its vapor and the equilibrium between a solid and its saturated solution.

5.2.1 Describing Chemical Equilibrium

Many chemical reactions are like the catalytic methanation reaction. Such reactions can be made to go predominantly in one direction or the other, depending on the conditions. Let's look more closely at this reversibility and see how to characterize its quantitatively.

When substances react, they eventually form a mixture of reactants and products in **dynamic equilibrium**. This dynamic equilibrium consists of a forward reaction, in which substance react to give products, and a reverse reaction, in which products react to give the original reactants. Both forward and reverse reactions occur at the same rate, or speed.

Consider the catalytic methanation reaction, which consists of forward and reverse reactions, as represented by the chemical equation

$$CO(g) + 3H_2(g) \rightleftharpoons CH_4(g) + H_2O(g)$$

Suppose you put 1.000 mol CO and 3.000 mol H_2 into a 10.00L vessel at 1200 K (927℃). The rate of the reactant CO and H_2 depends on the concentrations of CO and H_2. At first these concentrations are high, but as the substances react their concentrations decrease. The rate of the forward reaction, which depends on reactant concentrations, is large at first but steadily decreases. On the other hand, the concentrations of CH_4 and H_2O, which are zero at first, increase with time. The rate of the reverse reaction starts at zero and steadily increase. The forward rate decreases and the reverse rate increases until eventually the rates become equal. When that happens, CO and H_2 molecules are formed as fast as they react. The concentrations of reactants and products no longer change, and the reaction mixture has reached equilibrium. The amounts of substances in the reaction mixtures become constant when equilibrium is reached.

Chemical equilibrium is *the state reached by a reaction mixture when the rates of forward and reverse reactions have become equal*. If you observe the reaction mixture, you see no net change, although the forward and reverse reactions are continuing. The continuing forward and reverse make the equilibrium a dynamic process.

5.2.2 The Equilibrium Constant

When 1.000 mol CO and 3.000 mol H_2 react in a 10.00L vessel by catalytic methanation at 1200 K, they give an equilibrium mixture containing 0.613 mol CO, 1.839 mol H_2, 0.387 mol CH_4, and 0.387 mol H_2O. Let's call this Experiment 1. Now consider as similar experiment, Experiment 2, in which you start with an additional mole of CO. That is, you place 2.000 mol CO and 3.000 mol H_2 in a 10.00L vessel at 1200 K. At equilibrium, you find that the vessel contains 1.522 mol CO, 1.556 mol H_2, 0.478 mol H_2O. What you observe from the results of Experiments 1 and 2 is that the equilibrium composition depends on the amounts of starting substances. Nevertheless, you will see that all of the equilibrium compositions for a reaction at a given temperature are related by a quantity called the **equilibrium constant**.

Consider this reaction

$$eE + fF \rightleftharpoons gG + hH$$

where E, F, G, and H denote reactants and products, and e, f, g, and h are coefficients in the balanced chemical equation. The **equilibrium-constant expression** for a reaction is *an* expression obtained by multiplying the concentrations of products, dividing by the concentrations of reactants, and raising each concentration term to a power equal to the coefficient in the chemical equation. The **equilibrium constant K_c** is *the value obtained for the equilibrium-constant expression when equilibrium concentrations are substituted*. For the previous reaction, you have

$$K_c = \frac{[G]^g[H]^h}{[E]^e[F]^f}$$

Here you denote the molar concentration of a substance by writing its formula in square brackets. The subscript c on the equilibrium constant means that it is defined in terms of molar concentrations. The **law of mass action** is *a relation that states that the equilibrium-constant expression K_c are constant for a particular reaction at a given temperature, whatever equilibrium concentrations are substituted*.

In the gas-phase equilibria, it is often convenient to write the equilibrium constant in terms of partial pressure of gases rather than concentrations. Note that the concentration of a gas is proportional to its partial pressure at a fixed temperature. You can see this by looking at the ideal gas law, $PV=nRT$, and solving for n/V, which is the molar concentration of the gas. You get $n/V=P/RT$. In other words, the molar concentration of a gas equals its partial pressure divided by RT, which is constant at a given temperature.

When you express an equilibrium constant for a gaseous reaction in terms of partial pressure, you call it the **equilibrium constant K_p**. In general, the value of K_p is different from that of K_c. From the relationship for molar concentration $n/V=P/RT$, you can get $K_p = K_c(RT)^{\Delta n}$, where Δn is the sum of the coefficients of gaseous products in the chemical equation minus the sum of the coefficients of gaseous reactants.

5.2.3 Uses of the Equilibrium Constant

An equilibrium constant can be used to answer the following important questions:

(1) **Qualitatively interpreting the equilibrium constant.** By merely looking at the magnitude of K_c, you can tell whether a particular equilibrium favors products or reactants.

(2) **Predicting the direction of reaction.** Consider a reaction mixture that is not at equilibrium. By substituting the concentrations of substances that exist in a reaction mixture into an expression similar to the equilibrium constant and comparing with K_c, you can predict whether the reaction will proceed toward products or toward reactants (as defined by the way you write the chemical equation).

(3) **Calculating equilibrium concentrations.** Once you know the value of K_c for a reaction, you can determine the composition at equilibrium for any set of starting concentrations.

Obtaining the maximum amount of product from a reaction depends on the proper selection of reaction conditions. By changing these conditions, you can increase or decrease the yield of product. There are three ways to alter the equilibrium composition of a gaseous reaction mixture and possibly increase the yield of products:

(1) Changing the concentrations by removing products or adding reactants to the reaction vessel.

(2) Changing the partial pressure of gaseous reactants and products by changing the volume.

(3) Changing the temperature.

Note that a catalyst cannot alter equilibrium composition, although it can change the rate at which a product is formed.

Le Châtelier's Principle, which states that *when a system in chemical equilibrium is disturbed by a change of temperature, pressure, or a concentration, the system shifts in equilibrium composition in a way that tends to counteract this change of variable*, is useful in predicting the effect of changes mentioned above.

5.2.4 Applying Le Châtelier's Principle

5.2.4.1 *Changing the Concentration*

The reaction of yellow iron(Ⅲ) ions [Fe^{3+}(aq)] with colorless thiocyanate ions [SCN^-(aq)] produces deep red [$Fe(SCN)$]$^{2+}$(aq) ions. The reaction is reversible and an equilibrium is set up

$$Fe^{3+}(aq) + SCN^-(aq) \longrightarrow [Fe(SCN)]^{2+}(aq)$$

The intensity of the red color of the solution is a good indication of the concentration of [$Fe(SCN)$]$^{2+}$ ions in the mixture. If the concentration of either iron(Ⅲ) ions or thiocyanate ions is increased, the solution goes a darker red, showing the position of equilibrium has moved to the right. This agrees with the Le Châtelier's principle since moving to the right reduces the concentrations of the reactants and minimizes the effect of the imposed change. The concentration of iron(Ⅲ) ions can be reduced by adding ammonium chloride, as chloride ions react with iron(Ⅲ) ions to form [$FeCl_4$]$^-$ ions. By doing this, the red color of the solution becomes paler, indicating that the position of equilibrium has moved to the left.

5.2.4.2 *Changing the Pressure*

When a reversible reaction take place in the gas phase, increasing the pressure at which the reaction is carried out moves the position of equilibrium towards the side of the equation with fewer gas molecules. This reduces the pressure and minimizes the effect of the imposed change.

In industry, ammonia is made from nitrogen and hydrogen using Haber process.

$$N_2(g) + 3H_2(g) \Longleftrightarrow 2NH_3(g)$$

There are four molecules of gaseous reactants on the left-hand side of the equation, but only two molecules of gaseous product on the right-hand side. Thus, increasing the pressure causes the position of equilibrium to shift towards the right and increases the yield of ammonia. For this reason, Haber process is carried out at high pressure (25~150 atm, 1 atm = 101325 Pa).

5.2.4.3 *Changing the Temperature*

If the temperature is decreased, the position of equilibrium moves in the direction of the exothermic change because this increases the temperature and minimizes the effect of the imposed change.

For example, nitrogen dioxide (NO_2), a dark brown gas, is in equilibrium with dinitrogen tetroxide (N_2O_4), a colorless gas. The forward reaction is exothermic.

$$2NO_2(g) \rightleftharpoons N_2O_4(g) \qquad \Delta H^\ominus = -57 \text{ kJ/mol}$$

If a sealed glass container of the brown mixture is cooled in ice, the brown color of the mixture becomes colorless, because the position of equilibrium moves in the direction of the exothermic change towards the product.

New Words and Expressions

catalytic methanation n. 催化甲烷化
chemical equilibrium n. 化学平衡
dynamic equilibrium n. 动态平衡
equilibrium-constant expression n. 平衡常数表示式
equilibrium constant n. 平衡常数
law of mass action n. 质量作用定律
steam reforming n. 蒸汽重组法；蒸汽转化法
synthesis gas n. 合成气

Notes

1. It is a potentially useful reaction because it can be used to produce methane from coal or organic wastes from which carbon monoxide can be obtained.

参考译文：这是一个具有潜在应用价值的反应，因为它可以从煤或有机废物中获得一氧化碳进而产生甲烷。

2. Catalytic methanation is a reversible reaction, and depending on the reaction conditions, the final reaction mixture will have varying amounts of the products methane and steam, as well as the starting substances carbon monoxide and hydrogen.

参考译文：催化甲烷化反应是一个可逆的反应，使用不同的反应条件得到不同数量的产物（甲烷和蒸气）及起始物（一氧化碳和氢气）的混合物。

3. The equilibrium-constant expression for a reaction is an expression obtained by multiplying the concentrations of products, dividing by the concentrations of reactants, and raising each concentration term to a power equal to the coefficient in the chemical equation.

参考译文：反应的平衡常数表达式是将各产物的浓度相乘，然后除以各反应物的浓度，其中每个浓度项的幂为该物质在化学方程式中的系数。

4. The law of mass action is a relation that states that the of the equilibrium-constant expression K_c are constant for a particular reaction at a given temperature, whatever equilibrium concentrations are substituted.

参考译文：质量作用定律可以表述为对于特定的反应，在给定的温度下，无论平衡浓度是多少，平衡常数 K_c 都是常量。

5. Le Châtelier's Principle, which states that when a system in chemical equilibrium is disturbed by a change of temperature, pressure, or a concentration, the system shifts in equilibrium composition in a way that tends to counteract this change of variable, is useful in predicting the effect of changes mentioned above.

参考译文：勒夏特列原理指出，当化学系统处于平衡时，如果系统受到温度、压力或浓度变化的干扰，系统的转变方式将会抵消产生这些干扰的变量，这一原理在预测上述变化的影响是有用的。

Questions

1. What is chemical equilibrium?
2. Give an example for equilibrium-constant expression.
3. Are the equilibrium constants K_p and K_c the same for a gaseous reaction?

5.3 Thermodynamics

Urea, NH_2CONH_2, is an important industrial chemical. It is used to make synthetic resins for adhesives and melamine plastics. Its major use, however, is as a nitrogen fertilizer for plants. Urea is produced by reacting ammonia with carbon dioxide. We can write the overall reaction as:

$$2NH_3(g) + CO_2(g) \longrightarrow NH_2CONH_2(g) + H_2O(l)$$

Suppose you want to determine whether this or some other reactions could be useful for the industrial preparation of urea. Some of your immediate questions might be the following: does the reaction naturally go in the direction it is written? Will the reaction mixture contain a sufficient amount of the product at equilibrium? You may answer these questions by looking at the equilibrium constant. But now we want to discuss these questions form the point of view of thermodynamics.

Thermodynamics is *the study of the relationship between heat and other forms of energy involved in a chemical or physical process*. With only heat measurements of substances, you can answer the questions just posed. You can predict the natural direction of a chemical reaction, and you can also determine the composition of a reaction mixture at equilibrium. Consider, then, the possibility of reacting NH_3 and CO_2 to give urea. Just how do you apply thermodynamics to such a reaction?

5.3.1 First Law of Thermodynamics

Thermodynamics is described in terms of three laws. **The first law of thermodynamics** is essentially the law of conservation of energy applied to thermodynamic systems. The law of conservation of energy, you might recall, says that the total energy remains constant.

In thermodynamic system, the sum of the kinetic and potential energy of the particles making up the system is referred as its **internal energy**, U. The internal energy can change as the result of transfers of energy (or energy flows) into or out of the system, and the first law simply says that the change of internal energy equals the sum of these energy transfer.

These transfers, or flows of energy, are of two kinds: heat, q, and work, w. If a chemical reaction occurs in the system, **heat of reaction** may evolve or perhaps be absorbed. Energy as heat leaves or enters the system. The system may also do work on the surroundings or have work done on it by the surroundings. **Pressure-volume work** is a common form of work.

$$\text{Pressure-volume work} = -P\Delta V$$

Suppose the volume, V, of the system increases so that the system pushes against the external pressure of the atmosphere, P. Effectively the system lifts the atmosphere against gravity and does work on it. In doing this work, energy leaves the system (flows out of it) and goes into the surroundings. Or perhaps the volume of the system decreases so that the external pressure pushes against the system. In this case, work is done on the system; energy flows from the surroundings to the system.

According to the first law of thermodynamics, whenever a thermodynamic system undergoes a physical or chemical change, the change of internal energy, ΔU, of the system equals the sum of the heat, q, and work done, w, in that physical of chemical change.

$$\Delta U = q + w$$

When a physical or chemical change occurs against a constant external pressure, as would happen when the change occurs in a flask open to the atmosphere, it is useful to define the thermodynamic quantity call **enthalpy**, H. The enthalpy of a system is defined as the system's internal energy plus pressure time volume; it is a state function since U, P, and V are state functions. A **state function** is *a property of a system that depends only on its present state, which is determined by variables such as temperature and pressure.*

$$H = U + PV$$

With this definition, one can show the heat involved in a physical or chemical change for a given fixed pressure and temperature simply equals the change in enthalpy for the system.

$$q = \Delta H$$

It is useful to tabulate **standard enthalpies of formation**, $\Delta H_f^\ominus$, for substances. Using these values, you can calculate the standard enthalpy change for a reaction, which equals the heat of reaction.

$$\Delta H^\ominus = \sum n \Delta H_f^\ominus (\text{products}) - \sum m \Delta H_f^\ominus (\text{reactants})$$

Consider the reaction between NH_3 and CO_2 to produce urea and water

$$2NH_3(g) + CO_2(g) \longrightarrow NH_2CONH_2(g) + H_2O(l)$$

The standard enthalpies of formation at 25℃ for these substances are (kJ/mol): $NH_3(g)$, −45.9; $CO_2(g)$, −393.5; $NH_2CONH_2(g)$, −319.2; and $H_2O(l)$, −285.8. Substituting these values into the equation for the standard enthalpy change yields

$$\Delta H^\ominus = [(-319.2 - 285.8) - (-2 \times 45.9 - 393.5)] = -119.7 \text{ kJ/mol}$$

From the minus sign of $\Delta H^\ominus$, you can conclude that heat is evolved. The reaction is exothermic.

5.3.2 Second Law of Thermodynamics

Why does a chemical reaction go naturally in a particular direction? To answer this question, we need to look at spontaneous processes. A spontaneous process is a physical or chemical change that occurs by itself. It requires no continuing external force to make it happen. For example, a rock at the top of a hill rolls down; heat flows from a hot object to a cold one; an iron object rusts in moist air. These processes occur spontaneously, or naturally, without requiring an outside force. They continue until equilibrium is reached. If these processes were to go in the opposite direction, they would be nonspontaneous. They would require the application of an external force. The rolling of a rock uphill by itself is not a natural process; it is nonspontaneous. The rock would be moved to the top of the hill, but work would have to be expended. Heat can be made to flow from a cold to a hot object, but a heat pump or refrigerator is needed. Rust can be converted to iron, but the process requires chemical reactions used in the manufacture of iron from its ore (iron oxide).

When you ask whether a chemical reaction goes in the direction in which it is written, you are asking whether the reaction is spontaneous in this direction. The first law of thermodynamics cannot help you answer such a question. It does help you keep track of the various forms of energy in a chemical changing, using the constancy of total energy (conservation of energy). But although at one time it was though that spontaneous reactions must be exothermic ($\Delta H < 0$ kJ/mol), many spontaneous reactions are now known to be endothermic ($\Delta H > 0$ kJ/mol).

The second law of thermodynamics provides a way to answer questions about the spontaneity of a reaction. The second law is expressed in terms of a quantity called **entropy**.

Entropy, S, is a thermodynamic quantity that is a measure of how dispersed the energy of a system is among the different passible ways that system can contain energy. When the energy of a thermodynamic system is concentrated in a relative few energy states, the entropy of the system is low. When that same energy, however, is dispersed or spread out over a great many energy states, the entropy of the system is high. What you will see is that the entropy (energy dispersal) of a system plus its surroundings increases in a spontaneous process.

Let's look at some spontaneous process. Suppose you place a hot cup of tea on the table. Heat energy from the hot tea flows slowly to the table and to the air surrounding the cup. In this process, energy spreads out or disperses. The entropy of the system (tea cup) and its surroundings (table and surrounding air) increases in this spontaneous process.

As a slightly more complicated example, consider a rock at the top of a hill. The rock has potential energy relative to what it would have at the bottom of the hill. Imagine the rock rolling downhill. As it does so, its potential energy changes to kinetic energy, and during its descent atoms of the rock collide with those of the hillside and surrounding air, dispersing energy from the rock to its surroundings. The entropy of the system (rock) plus its surroundings (hillside and surrounding air) increases in this spontaneous process.

As a final example, imagine a flask containing a gas connected to an evacuated flask by a valve or stopcock. When the valve is opened, gas in the flask flows into the space of the evacuated flask. In this case, the kinetic energy of the gas molecules spreads out or disperses over the volumes of both flasks. The entropy of this system (both flasks) increases in this spontaneous process. In this case, we assume that the surroundings do not participate in the overall change (there is no entropy change in the surroundings); the entropy change occurs entirely within the system.

In each of these spontaneous processes, energy has been dispersed, or spread out. The entropy of the system plus surroundings has increased. We can state this precisely in terms of the **second law of thermodynamics**, which states that *the total entropy of a system and its surroundings always increases for a spontaneous process*. Note that entropy (energy dispersal) is quite different from energy itself. Energy can be neither created not destroyed during a spontaneous, or natural, process—its total amount remains fixed. But energy is dispersed in a spontaneous process, which means that entropy is produced (or created) during such a process.

For a spontaneous process carried out at a given temperature, the second law can be restated in a form that refers only to the system (and not to the system plus surroundings, as in the previous statement of the second law). We will find this new statement particularly useful in analyzing chemical problems.

Suppose a spontaneous process occur within a system that is in thermal contact with its surroundings at a given temperature T—say, a chemical reaction in a flask. As the chemical reaction occurs, entropy is produced (or created) within the system (the flask) a result of this spontaneous process. At the same time, heat might flow into or out of the system as a result of the thermal contact. (In other words, the reaction may be endothermic or exothermic.) Heat flow is also a flow of entropy, because it is a dispersal of energy, either into the flask or outside of it. In general, the entropy change associated with a flow of heat q at an absolute temperature T can be shown to equal q/T. The net change of entropy in the system (in the flask) is the entropy created during the spontaneous chemical reaction that occurs plus the entropy change that is associated with the heat flow (entropy flow).

$$\Delta S = \text{entropy created} + \frac{q}{T}$$

It might be helpful to consider an analogy. A man makes (creates) candies at his store (the system). He also buys candies from a wholesaler and sells candies at his store. The change in number of candies in the store in any given interval of time equals the number made (number created) plus the number of bought minus the number sold (flow of candies into or out of his store). We obtain an equation similar to the one we wrote for the entropy change ΔS.

Changes of candies in store = candies created + flow of candies into or out of store

However, our analogy breaks down if someone eats candies in the store. Whereas entropy can only be created, candies can be created and destroyed (eaten).

The quantity of entropy created during a spontaneous process is a positive quantity—the entropy increases as it is created. If we delete "entropy created" from the right side of the equation for ΔS, we know that the left side is then greater than the right side:

$$\Delta S > \frac{q}{T}$$

We can now restate **the second law of thermodynamics** as follows: *For a spontaneous process at a given temperature T, the change in entropy of the system is greater than the heat divided by the absolute temperature, q/T.*

5.3.3 Third Law of Thermodynamics

To determine experimentally the entropy of a substance, you first measure the heat absorbed by the substance by warming it at various temperature. That is, you find the heat capacity at different temperatures. You then calculate the entropy as we will describe. This determination of the entropy is based on the third law of thermodynamics.

The **third law of Thermodynamics** states that a substance is perfectly crystalline at 0 K has an entropy of zero. This seems reasonable. A perfectly crystalline substance at 0 K should have perfect order. When the temperature is raised, however, the substance increases in entropy as it absorbs heat and energy disperses through it.

You can determine the entropy of a substance at a temperature other than 0 K—say, at 298 K (about 25℃)—by slowly heating the substance from near 0 K to 298 K. Recall that the entropy change ΔS that occurs when heat is absorbed at a temperature T is q/T. Suppose you heat a substance from 0 K to 2 K, and the heat absorbed is 0.19 J. You find the entropy change by diving the heat absorbed by the average absolute temperature [(0.0+2.0)/2 K=1.0 K]. Therefore, ΔS equals 0.19 J/1.0 K = 0.19 J/K. This gives you the entropy of the substance at 2.0 K. Now you heat the substance from 2.0 K to 4.0 K, and this time 0.88 J of heat is absorbed. The average temperature is (2.0+4.0)/2 K=3.0 K, and the entropy change is 0.88 J/3.0 K≈0.29 J/K. The entropy of the substance at 4.0 K is (0.19+0.29) J/K = 0.48 J/K. Proceeding this way, you can eventually get the entropy at 298 K.

The entropy of a substance changes with temperature. Note that the entropy increases gradually as the temperature increases. But when there is a phase change (for example, from solid to liquid), the entropy increases sharply. The entropy change for the phase transition is calculated from the enthalpy of the phase transition. The **standard entropy** of a substance or an ion, also called its **absolute entropy**, $S^{\ominus}$, is *the entropy value for the standard state of the species* (indicated by the superscript degree sign). For a substance, the standard state is the pure substance at 1 atm pressure. For a species in solution, the standard state is the 1 mol/L solution. Note that elements have nonzero values, unlike standard enthalpies of formation, $\Delta H_f^{\ominus}$, which by convention are defined to be zero. The symbol $S^{\ominus}$, rather than

$\Delta S^{\ominus}$, is chosen for standard entropies to emphasize that they originate from the third law.

New Words and Expressions

adhesive /əd'hi:sɪv/ n. 黏合剂
descent /dɪ'sent/ n. 下降
endothermic /ˌendəʊ'θɜ:mɪk/ adj. 吸热的
enthalpy /en'θælpɪ/ n. 焓
entropy /'entrəpi/ n. 熵
exothermic /ˌeksəʊ'θɜ:mɪk/ adj. 放热的
heat of reaction n. 反应热
internal energy n. 内能

melamine /'meləmi:n/ n. 三聚氰胺
resin /'rezɪn/ n. 树脂
standard enthalpies of formation n. 标准生成焓
state function n. 状态函数
stopcock /'stɒpkɒk/ n. 旋塞，活栓
valve /vælv/ n. 阀门

Notes

1. exo-表示"外部，外面，在……外"，如 exothermic（放热的），exosphere（外大气层），exotic（外来的）。

2. endo-表示"内部"，如 endothermic（吸热的），endogenous（内生的），endocardial（心脏内的）。

3. When a physical or chemical change occurs against a constant external pressure, as would happen when the change occurs in a flask open to the atmosphere, it is useful to define the thermodynamic quantity call enthalpy, H.

参考译文：当在恒定的外部压力下发生物理或化学变化时，就像在向大气敞开的烧瓶中发生变化一样，需要定义一个称为焓（H）的热力学量。

Questions

1. What are the three laws of thermodynamics?
2. Give two examples of spontaneous and nonspontaneous processes, respectively.

5.4 Solutions

When sodium chloride dissolves in water, the resulting uniform dispersion of ions in water is called a solution. In general, a **solution** is *a homogeneous mixture of two or more substances, consisting of ions or molecules*. A **colloid** is similar in that it appears to be homogeneous like a solution. In fact, it consists of comparatively large particles of one substance dispersed throughout another substance or solution.

5.4.1 Types of Solutions

Solution may exist in any of the three states of matter; that is, they may be gases, liquids, or solids. Some examples are listed in Table 5.1. The terms solute and solvent refer to the components of a solution. The **solute**, in the case of a solution of a gas or solid dissolved in a liquid, is the gas or solid; in other cases, the solute is the component in smaller amount. The **solvent**, in a solution of a gas or solid dissolved in a liquid, is the liquid; in other cases, the solvent is the component in greater amount. Thus, when sodium chloride is dissolved in water, we have created a solution in which sodium chloride is the solute and water is the solvent.

Table 5.1 Examples of solutions

Solute	Solvent	State of Resulting solution	Example
Gas	Gas	Gas	Air
Gas	Liquid	Liquid	Soda water
Gas	Solid	Solid	H_2 gas in palladium
Liquid	Liquid	Liquid	Acetone in water
Liquid	Solid	Solid	Mercury in silver
Solid	Liquid	Liquid	Brine
Solid	Solid	Solid	Gold-silver alloy

(1) Gaseous Solutions. In general, nonreactive gases or vapors can mix in all proportions to give a gaseous mixture. *Fluids that mix with or dissolve in each other in all proportions* are said to be **miscible fluids**. Gases are thus miscible. (If two fluids do not mix but, rather, form two layers, they are said to be **immiscible**.) Air, which is a mixture of nitrogen, oxygen, and smaller amounts of other gases, is an example of a gaseous solution.

(2) Liquid Solutions. Most liquid solutions are obtained by dissolving a gas, liquid, or solid in some liquid. Soda water, for example, consists of a solution of carbon dioxide gas in water. Acetone, C_3H_6O, in water is an example of a liquid-liquid solution. Brine is water with sodium (a solid) dissolved in it. Seawater contains both dissolved gases (from air) and solid (mostly sodium chloride).

It is also possible to make a liquid solution by mixing two solids together. Consider a potassium-sodium alloy. Both potassium and sodium are solid at room temperature, but a liquid solution results when the mixture contains 10% to 50% sodium.

(3) Solid Solutions. Solid solutions are also possible. Dental-filling alloy is a solution of mercury (a liquid) in silver (a solid), with small amounts of other metals.

5.4.2 Solubility

The amount of substance that will dissolve in a solvent depends on both the substance and the solvent. We describe the amount that dissolves in terms of solubility.

To understand the concept of solubility, consider the process of dissolving sodium chloride in water. Sodium chloride is an ionic substance, and it dissolves in water as Na^+ and Cl^- ions. If you could view the dissolving of sodium chloride at the level of ions, you would see a dynamic process. Suppose you stir 50.0 g of sodium chloride crystals into 100 mL of water at 20℃. Sodium ions and chloride ions leave the surface of the crystals and enter the solution. The ions move about at random in the solution and may by chance collide with a crystal and stick, thus returning to the crystalline state. As the sodium chloride continues to dissolve, more ions enter the solution, and the rate at which they return to the crystalline state increases (the more ions in solution, the more likely ions are to collide with the crystals and stick). Eventually, a dynamic equilibrium is reached in which the rate at which ions leave the crystals equals the rate at which ions return to the crystals. You write the dynamic equilibrium this way

$$NaCl(s) \rightleftharpoons Na^+(aq) + Cl^-(aq)$$

At equilibrium no more sodium chloride appears to dissolve; 36.0 g has gone into solution, leaving 14.0 g of crystals at the bottom of the vessel. You have a **saturated solution**—*a solution that is in equilibrium with respect to a given dissolved substance*. The solution is saturated with respect to NaCl, and no more NaCl can dissolve. The **solubility** of sodium chloride in water (*the amount that dissolves in a given quantity of water at a given temperature to give a saturated solution*) is 36.0 g/100 mL at 20℃. Note that if you had mixed 30.0 g of sodium chloride with 100 mL of water at 20℃, all of the crystals would have dissolved. You would have an **unsaturated solution**, *a solution not in equilibrium with respect to a given dissolved substance and in which more of the substance can dissolve.*

Sometimes it is possible to obtain a **supersaturated solution**, *a solution that contains more dissolved substance than a saturated solution does*. For example, the solubility of sodium thiosulfate, $Na_2S_2O_3$, in water at 100℃ is 231 g/100 mL. But at room temperature, the solubility is much less—about 50 g/100 mL. Suppose you prepare a solution saturated with sodium thiosulfate at 100℃. You might expect that as the water solution was cooled, sodium thiosulfate would crystallize out. In fact, if the solution is slowly cooled to room temperature, this does not occur. Instead, the result is a solution in which 231 g of sodium thiosulfate is dissolved in 100 mL of cool water, compared with the 50 g you would normally expect to find dissolved.

Supersaturated solutions are not in equilibrium with the solid substance. If a small crystal of sodium thiosulfate is added to a supersaturated solution, the excess immediately crystalizes out. Crystallization from a supersaturated solution is usually quite fast and dramatic.

5.4.3 Molecular Solutions

The simplest example of a molecular solution is one gas in another gas. Air, essentially a solution of nitrogen and oxygen, is an example. The intermolecular forces in gases are weak. The only solubility factor of importance is the tendency for molecules to mix. Gases are therefore miscible.

Substances may be miscible even when the intermolecular forces are not negligible. Consider the solution of the two similar liquid hydrocarbons heptane (C_7H_{16}) and octane (C_8H_{18}) which are components of gasoline. The intermolecular attractions are due to London forces, and those between heptane and octane molecules are nearly equal to those between octane and octane molecules and heptane and heptane molecules. The different molecular attractions are about the same strength, so there are no favored attractions. Octane and heptane molecules tend to move freely through one another. Therefore, the tendency of molecules is to mix results in miscibility of the substances.

As a counterexample, consider the mixing of heptane with water. There are strong bonding forces between water molecules. For heptane to mix with water, hydrogen bonds must be broken and replaced by much weaker London forces between water and heptane. In this case, the maximum forces of attraction among molecules (and therefore the lower energy) result if the heptane and water remain unmixed. Therefore, heptane and water are nearly immiscible.

The statement "like dissolve like" succinctly expresses these observations. That is, substances with similar intermolecular attractions are usually soluble in one another. The two similar hydrocarbons heptane and octane are completely miscible, whereas octane and water (with dissimilar intermolecular attractions) are immiscible.

For a series of alcohols (organic, or carbon-containing, compounds with an —OH group) listed in Table 5.2, the solubility in water decreases from miscible to slightly soluble. Water and alcohols are alike in having —OH groups through which strong hydrogen bonding attraction arise.

Table 5.2 Solubilities of alcohols in water

Name	Solubility in H_2O (g/100 g H_2O at 20℃)
Methanol	Miscible
Ethanol	Miscible
1-Propanol	Miscible
1-Butanol	7.9
1-Pentanol	2.7
1-Hexanol	0.6

The attraction between a methanol molecule (CH_3OH) and a water molecule is nearly as strong as that between two methanol molecules or between two water molecules. Methanol and water are miscible, as are ethanol and water as well as 1-propanol and water. However, as the hydrocarbon end (R—) of the alcohol becomes the more prominent portion of the molecule, the alcohol becomes less like water. Now, the force of attraction between alcohol and water molecules are weaker than those between two alcohol molecules or between two water molecules. Therefore, the solubility of alcohols decreases with increasing length of R.

5.4.4　Ionic Solutions

Ionic substances differ markedly in their solubilities in water. For example, sodium chloride, NaCl, has a solubility of 36 g per 100 mL of water at room temperature, whereas calcium phosphate, $Ca_3(PO_4)_2$, has a solubility of only 0.002 g per 100 mL of water at room temperature. In most cases, these differences in solubility can be explained in terms of the different energies of attraction between ions in the crystal and between ions and water.

The energy of attraction between an ion and a water molecule is due to an **ion-dipole force**. Water molecules are polar, so they tend to orient with respect to nearby ions. In the case of a positive ion (Na^+, for example), water molecules orient with their oxygen atoms (the negative ends of the molecule dipoles) toward the ion. In the case of negative ion (for instance, Cl^-), water molecules orient with their hydrogen atoms (the positive ends of the molecular dipoles) toward the ion.

The attraction of ions for water molecules is called **hydration**. Hydration of ions favors the dissolving of an ionic solid in water. As ions on the surface become partially hydrated, bonds to the crystal weaken to the point the ions break free and subsequently become completely hydrated in solution.

If the hydration of ions were the only factor in the solution process, you would expect all ionic solids to be soluble in water. The ions in a crystal, however, are very strongly attracted to one another. Therefore, the solubility of an ionic solid depends not only on the energy of hydration of ions (energy associated with the attraction between ions and water molecules) but also on **lattice energy**, the energy holding ions together in the crystal lattice. Lattice energy works against the solution process, so an ionic solid with relatively large lattice energy is usually insoluble.

Lattice energies depend on the charges on the ions as well as on the distance between the centers of neighboring positive and negative ions. The greater the magnitude of ion charges, the greater is the lattice energy. For this reason, you might expect substances with singly charged ions to be comparatively soluble and those with multiply charged ions to be less soluble. This is borne out by the fact that compounds of the alkali metal ions (such as Na^+ and K^+) and ammonium ions (NH_4^+) are generally soluble, whereas those of phosphate ions (PO_4^{3-}) combined with multiply-charged cations, for example, are generally insoluble.

Lattice energy is also inversely proportional to the distance between neighboring ions, and this distance depends on the sum of the radii of the ions. For example, the lattice energy of magnesium hydroxide, $Mg(OH)_2$, is inversely proportional to the sum of the radii of Mg^{2+} and OH^-. In the series of alkaline earth hydroxides—$Mg(OH)_2$, $Ca(OH)_2$, $Sr(OH)_2$, and $Ba(OH)_2$—the lattice energy decreases as the radius of the alkaline earth ion increases (from Mg^{2+} to Ba^{2+}). If the lattice energy alone determines the trend in solubilities, you should expect the solubility to increase from magnesium hydroxide to barium hydroxide: the expected solubility ranking being $Mg(OH)_2 < Ca(OH)_2 < Sr(OH)_2 < Ba(OH)_2$. In fact, this is what you find. Magnesium hydroxide is insoluble in water, and barium hydroxide is soluble. But this is not the whole story. The energy of hydration also depends on ionic radius. A small ion has a concentrated electric charge and a strong electric field that attracts water molecules. Therefore, the hydration is greatest for a small ion such as Mg^{2+} and least for a large ion such as Ba^{2+}. If the energy of hydration of ions alone determined the trend in solubilities, you would expect the solubilities to decrease from magnesium hydroxide to barium hydroxide, rather than to increase. (We should also add that the energy of hydration increases with the charge on the ion; energy of hydration is greater for Mg^{2+} than for Na^+.)

The explanation for the observed solubility trend in the alkaline earth hydroxides is that the lattice energy decreases more rapidly in the series of $Mg(OH)_2$, $Ca(OH)_2$, $Sr(OH)_2$, and $Ba(OH)_2$ than does the energy of hydration in the series of ions Mg^{2+}, Ca^{2+}, Sr^{2+}, and Ba^{2+}. For this reason, the lattice-energy factor dominates this solubility trend.

You see the opposite solubility trend when the energy of hydration decreases more rapidly so that it dominates the trend. Let us consider the alkaline earth sulfates. Here the lattice energy depends on the sum of the radius of the cation and the radius of the sulfate ions. Because the sulfate ion, SO_4^{2-}, is much larger than the hydroxide ion, OH^-, the precent change in the lattice energy in going through the series of sulfate from $MgSO_4$ to $BaSO_4$ is smaller than in the hydroxide. The lattice energy changes less, and the energy of hydration of the cation decreases by a great amount. Now the energy of hydration dominates the solubility trend, and the solubility decreases from magnesium sulfate to barium sulfate. Magnesium sulfate is soluble in water, and barium sulfate is insoluble.

New Words and Expressions

colloid /ˈkɒlɔɪd/ n. 胶体
homogeneous /ˌhɒməˈdʒiːniəs/ adj. 均相的
hydration /haɪˈdreɪʃn/ n. 水合
immiscible /ɪˈmɪsəbl/ adj. 不互溶的
ion-dipole force n. 离子-偶极力
lattice energy n. 晶格能
miscible /ˈmɪsəbl/ adj. 可溶混的

saturated solution n. 饱和溶液
solubility /ˌsɒljʊˈbɪləti/ n. 溶解度
solute /ˈsɒljuːt/ n. 溶质
solution /səˈluːʃn/ n. 溶液
solvent /ˈsɒlvənt/ n. 溶剂
supersaturated solution n. 过饱和溶液
unsaturated solution n. 不饱和溶液

Notes

1. im-有多种含义，其中一种表示"不，无"，如 immiscible（不互溶的），impure（不纯的），impenetrable（不能穿透的），impossible（不可能的）。

2. hydr-表示"水"，如 hydrate（水合物），hydration（水合），carbohydrate（碳水化合物），dehydrate（脱水），anhydrous（无水的），hydrant（消火栓）。

3. As the sodium chloride continues to dissolve, more ions enter the solution, and the

rate at which they return to the crystalline state increases (the more ions in solution, the more likely ions are to collide with the crystals and stick).

参考译文：当氯化钠继续溶解时，更多的离子进入溶液，并且它们回到结晶状态的速率增加（溶液中的离子越多，离子与晶体碰撞并黏附的可能性越大）。

4. The intermolecular attractions are due to London forces, and those between heptane and octane molecules are nearly equal to those between octane and octane molecules and heptane and heptane molecules.

参考译文：分子间吸引力来自伦敦力，所以庚烷和辛烷分子之间的吸引力几乎等于辛烷分子之间的吸引力。

Questions

1. Which of the following ions would be expected to have the largest hydration energy, Na^+ or K^+?

2. Would naphthalene be more soluble in methanol or in benzene?

5.5 Catalysis

A catalyst is a substance that has the seemingly miraculous power of speeding up a reaction without being consumed by it. In theory, you could add a catalyst to a reaction mixture and, after the reaction, separate that catalyst and use it over and over again. In practice, there is often some loss of catalyst through other reactions that can occur at the same time. **Catalysis** is the increase in rate of a reaction that results from the addition of a catalyst.

Catalysts are of enormous important to the reaction industry, because they allow a reaction to occur with a reasonable rate at a much lower temperature than otherwise; lower temperatures translate into lower energy costs. Moreover, catalysts are often quite specific—they increase the rate of certain reaction, but not others. For instance, an industry chemistry can start with a mixture of carbon monoxide and hydrogen and produce methane gas using one catalyst or produce gasoline using another catalyst. The most remarkable catalysts are enzymes. Enzymes are the marvelously selective catalysts employed by biological organisms. A biological cell contains thousands of different enzymes that in effect direct all of the chemical processes that occur in the cell.

How can we explain how a catalyst can influence a reaction without being consumed by it? Briefly, the catalyst must participate in at least one step of a reaction and be regenerated in a later step. Consider the commercial preparation of sulfuric acid (H_2SO_4) from sulfur dioxide (SO_2). The first step involves the reaction of SO_2 with O_2 to produce sulfur trioxide (SO_3). For this reaction to occur at an economical rate, it requires a catalyst. An early industrial process employed nitrogen monoxide, NO, as the catalyst.

$$2SO_2(g) + O_2(g) \xrightarrow{NO} 2SO_3(g)$$

Nitrogen monoxide does not appear in the overall equation but must participate in the reaction mechanism. Here is a proposed mechanism

$$2NO + O_2 \longrightarrow 2NO_2$$
$$NO_2 + SO_2 \longrightarrow NO + SO_3$$

To obtain the overall reaction, the last step must occur twice each time the first step occurs once. As you can see from the mechanism, two molecules of NO are used up in the first step and are regenerated in the second step.

Note that the catalyst is an active participant in the reaction. But how does this participation explain the increase in speed of the catalyzed reaction over the uncatalyzed reaction? The Arrhenius equation provides an answer. The catalyzed reaction mechanism makes available a reaction path having an increased overall rate of reaction. It increases this rate either by increasing the frequency factor A or by decreasing the activation energy E_a. The most dramatic effect comes from decreasing the activation energy, because it occurs as an exponent in the Arrhenius equation.

The depletion of ozone in the stratosphere by Cl atoms provides an example of the lowering of activation energy by a catalyst. Ozone is normally present in the stratosphere and provides protection against biologically destructive, short-wavelength ultraviolet radiation from the sun. Some recent ozone depletion in the stratosphere is believed to result from the Cl-catalyzed decomposition of O_3. Cl atoms in the stratosphere originate from the decomposition of chlorofluorocarbons (CFCs), which are compounds manufactured as refrigerants, aerosol propellants, and so forth. These Cl atoms react with ozone to form ClO and O_2, and the ClO reacts with O atoms (normally in the stratosphere) to produce Cl and O_2.

Step 1: $Cl(g) + O_3(g) \longrightarrow ClO(g) + O_2(g)$

Step 2: $ClO(g) + O(g) \longrightarrow Cl(g) + O_2(g)$

Overall: $O_3(g) + O(g) \longrightarrow 2O_2(g)$

The net result is the decomposition of ozone with O atoms to produce O_2. The uncatalyzed reaction has such a large activation energy that its rate is extremely low. The addition of chlorine atoms provides an alternative pathway to the same overall reaction, but has much lower activation energy and therefore an increased rate.

5.5.1 Homogeneous Catalysis

The oxidation of sulfur dioxide using nitrogen monoxide as a catalyst is an example of **homogeneous catalysis**, which is *the use of catalyst in the same phase as the reacting species*. The catalyst NO and the reacting species SO_2 and O_2 are all in the gas phase. Another example occurs in the oxidation of thallium(Ⅰ) to thallium(Ⅲ) by cerium(Ⅳ) in aqueous solution.

$$2Ce^{4+}(aq) + Ti^+(aq) \longrightarrow 2Ce^{3+}(aq) + Ti^{3+}(aq)$$

The uncatalyzed reaction is very slow; presumably it involves the collision of three positive ions. The reaction can be catalyzed by manganese(II) ion, however. The mechanism is thought to be

$$Ce^{4+}(aq) + Mn^{2+}(aq) \longrightarrow Ce^{3+}(aq) + Mn^{3+}(aq)$$
$$Ce^{4+}(aq) + Mn^{3+}(aq) \longrightarrow Ce^{3+}(aq) + Mn^{4+}(aq)$$
$$Mn^{4+}(aq) + Ti^{+}(aq) \longrightarrow Mn^{2+}(aq) + Ti^{3+}(aq)$$

Each step is bimolecular.

A striking example of a homogeneous catalyst increasing the rate of a reaction is the decomposition of aqueous hydrogen peroxide to form water and oxygen.

$$2H_2O_2(aq) \longrightarrow O_2(g) + 2H_2O(l)$$

Under standard conditions without a catalyst, this reaction is very slow, so slow in fact that when you observe a solution of hydrogen peroxide, you cannot visibly detect the reaction. However, when a solution containing aqueous potassium iodide is added to hydrogen peroxide, the rate of the reaction increases dramatically. In this reaction, the aqueous iodide anion is the homogeneous catalyst. The proposed mechanism for the most important reaction in the solution is

Step 1: $H_2O_2(aq) + I^-(aq) \longrightarrow IO^-(aq) + 2H_2O(l)$ (slow)

Step 2: $IO^-(aq) + H_2O_2(aq) \longrightarrow H_2O(l) + O_2(g) + I^-(aq)$ (fast)

Overall: $2H_2O_2(aq) \longrightarrow O_2(g) + 2H_2O(l)$

5.5.2 Heterogeneous Catalysis

Some of the most important industrial reactions involve **heterogeneous catalysis**—that is, *the use of a catalyst that exists in a different phase from the reacting species, usually a solid catalyst in contact with a gaseous or liquid solution of reactants.* Such surface, or heterogeneous, catalysis is thought to occur by chemical adsorption of the reactant onto the surface of the catalyst. **Adsorption** is the attraction of molecules to a surface. In **physical adsorption**, the attraction is provided by weak intermolecular forces. **Chemisorption**, by contrast, is the binding of a species to a surface by chemical binding forces. It may happen that bonds in the species are broken during chemisorption, and this may provide the basis of catalytic cation in certain cases.

An example of heterogeneous catalysts involving chemisorption is provided by catalytic hydrogenation. This is the addition of H_2 to a compound, such as one with a carbon-carbon double bond, using a catalyst of platinum or nickel metal. Vegetable oils, which contain carbon-carbon double bonds, are changed to solid fat (shortening) when the bonds are catalytically hydrogenated. In the case of ethylene, C_2H_4, the equation is

$$\underset{\text{ethylene}}{\overset{H}{\underset{H}{>}}C=C\overset{H}{\underset{H}{<}}} + H_2 \xrightarrow{\text{Pt catalyst}} \underset{\text{ethane}}{H-\overset{H}{\underset{H}{C}}-\overset{H}{\underset{H}{C}}-H}$$

A mechanism for this reaction is represented by four steps: (a) ethylene and hydrogen molecules diffuse to the catalyst surface, (b) the molecules form bonds to the catalyst surface (the H_2 molecules dissociate to atoms in the process), (c) H atoms migrate to the C_2H_4 molecule, where they react to form C_2H_6, (d) C_2H_6 diffuse away from the catalyst.

Surface catalysts are used in the catalytic converters of automobiles to convert substances that would be atmospheric pollutants, such as CO and NO, into harmless substances, such as CO_2 and N_2.

5.5.3 Enzyme Catalysis

Almost all enzymes, the catalysts of biological organisms, are protein molecules with molecular weights ranging to over a million amu. An enzyme has enormous catalytic activity, converting a thousand or so reactant molecules to products in a second. Enzymes are also highly specific, each enzyme acting only on a specific substance, or a specific type of substance, catalyzing it to undergo a particular reaction. For example, the enzyme sucrase, which is present in the digestive fluid of the small intestine, catalyzes the reaction of sucrose (table sugar) with water to form the simpler sugars glucose and fructose. *The substance whose reaction the enzyme catalyzes* is called the **substrate**. Thus, the enzyme sucrase acts on the substrate sucrose.

The enzyme molecule is a protein chain that tends to fold into a roughly spherical form with an active site at which the substrate molecule binds and the catalysis takes place. The substrate molecule, S, fits into the active site on the enzyme molecule, E, somewhat in the way a key fits into a lock, forming an enzyme-substrate complex, ES. (The lock-and-key model is only a rough approximation, because the active site on an enzyme deforms somewhat to fit the substrate molecule.) In effect, the active site "recognizes" the substrate and gives the enzyme its specificity. On binding to the enzyme, the substrate may have bonds weakened or new bonds formed that help yield the product, P.

$$E + S \longrightarrow ES \longrightarrow E + P$$

The formation of the enzyme-substrate complex, ES, provides a new pathway to product with a lower activation energy, which increases the reaction rate.

New Words and Expressions

adsorption /ədˈzɔːpʃ(ə)n/ n. 吸附
catalysis /kəˈtæləsɪs/ n. 催化
catalyst /ˈkætəlɪst/ n. 催化剂
chemisorption /kemɪˈzɔrpʃən/ n. 化学吸附
depletion /dɪˈpliːʃn/ n. 耗尽
digestive fluid n. 消化液
heterogeneous catalysis n. 多相催化

homogeneous catalysis n. 均相催化
intestine /ɪnˈtestɪn/ n. 肠
shortening /ˈʃɔːtnɪŋ/ n. （制作油酥点心用的）起酥油
stratosphere /ˈstrætəsfɪə(r)/ n. 平流层
sucrase /ˈs(j)uːkreɪz/ n. 蔗糖酶

Notes

1. 其他种类的催化：photocatalysis（光催化），electrocatalysis（电催化），biocatalysis（生物催化）。

2. The oxidation of sulfur dioxide using nitrogen monoxide as a catalyst is an example of homogeneous catalysis, which is the use of catalyst in the same phase as the reacting species.

参考译文：使用一氧化二氮作为催化剂的二氧化硫氧化是均相催化的一个例子，即使用与反应物处于同一相的催化剂。

3. Some of the most important industrial reactions involve heterogeneous catalysis—that is, the use of a catalyst that exists in a different phase from the reacting species, usually a solid catalyst in contact with a gaseous or liquid solution of reactants.

参考译文：多相催化被用于一些重要的工业反应，即使用与反应物不同相的催化剂，通常是与气态或液态的反应物溶液接触的固体催化剂。

Questions

1. Give the definition of a catalyst.
2. Give examples for homogeneous catalysis and heterogeneous catalysis, respectively.

5.6 Electrochemical Cells

An **electrochemical cell** is *a system consisting of electrodes that dip into an electrolyte and in which a chemical reaction either uses or generates an electric current*. A **galvanic cell** (or **voltaic cell**) is an electrochemical cell in which a spontaneous reaction generates an electric current. An **electrolytic cell** is an electrochemical cell in which an external energy source drives an otherwise nonspontaneous reaction.

5.6.1 Galvanic Cells

When zinc metal is placed in a solution containing copper(Ⅱ) ions. Zn is oxidized to Zn^{2+} whereas Cu^{2+} ions are reduced to Cu.

$$Zn(s) + Cu^{2+}(aq) \longrightarrow Zn^{2+}(aq) + Cu(s)$$

The electrons are transferred directly from the reducing agent, Zn, to the oxidizing agent, Cu^{2+}, in solution. However, if we physically separate two half-reaction from each other, we can arrange it such that the electrons must travel through a wire to pass from the Zn atom to the Cu^{2+} ions. As the reaction progresses, it generates a flow of electrons through the wire and thereby generates electricity.

The experiment apparatus for generating electricity through the use of a spontaneous reaction is called a **galvanic cell**. A zinc bar is immersed in an aqueous $ZnSO_4$ solution in one container, and a copper bar is immersed in an aqueous $CuSO_4$ solution in another container. The cell operates on the principle that the oxidation of Zn to Zn^{2+} and the reduction of Cu^{2+} to Cu can be made to take place simultaneously in separate locations with the transfer of electrons between them occurring through an external wire. The zinc and copper bars are called **electrodes**. By definition, the **anode** in a galvanic cell is the electrode at which oxidation occurs and the **cathode** is the electrode at which reduction occurs. (Each combination of container, electrode, and solution is called a **half-cell**.)

The half-reactions for this galvanic cell are

Oxidation: $\quad\quad\quad\quad Zn(s) \longrightarrow Zn^{2+}(aq) + 2e^-$

Reduction: $\quad\quad\quad\quad Cu^{2+}(aq) + 2e^- \longrightarrow Cu(s)$

To complete the electric circuit, and allow electrons to flow through the external wire, the solutions must be connected by a conducting medium through which the cations and anions can move from one half-cell to the other. This requirement is satisfied be a **salt bridge**, which is an inverted U tube of an electrolyte (such as KCl or NH_4NO_3) in a gel that is connected to the two half-cells of a voltaic cell; the salt bridge allows the flow of ions but prevents the mixing of the different solution that would allow direct reaction of the cell reactants. The ions in the salt bridge must not react with the other ions in solution or with the electrodes. During the course of the redox reaction, electrons flow through the external wire from the anode (Zn electrode) to the cathode (Cu electrode). In the solution, the cations (Zn^{2+}, Cu^{2+}, and K^+) move toward to cathode, while the anions (SO_4^{2-} and Cl^-) move toward the anode. Without the salt bridge connecting the two solutions, the buildup of positive charge in the anode compartment (due to the departure of electrons and the resulting formation of Zn^{2+}) and the buildup of negative charge in the cathode compartment (created by the arrival of electrons and the reduction of Cu^{2+} to Cu) would quickly prevent the cell from operating.

An electric current flows from the anode to the cathode because there is a difference in electrical potential energy between the electrodes. This flow of electric current is analogous to the flow of water down a waterfall, which occurs because there is difference in the gravitational potential energy, or the flow of gas from a high-pressure region to a low-pressure region. Experimentally the difference in electrical potential between the anode and the cathode is measured by a voltmeter and the reading (in volts) is called the **cell potential** (E_{cell}). The potential of a cell depends not only on the nature of the electrodes and the ions in solution, but also on the concentrations of the ions and the temperature at which the cell is operated.

The conventional notation for representing galvanic cells is the **cell diagram**. For the cell discussed in this text, if we assume that the concentrations of and ions are 1 mol/L, the cell diagram is

$$Zn(s) | Zn^{2+}(1\ mol/L) \| Cu^{2+}(1\ mol/L) | Cu(s)$$

The single vertical line represents a phase boundary. For example, the zinc electrode is a solid and the Zn^{2+} ions are in solution. Thus, we draw a line between Zn and Zn^{2+} to show the phase boundary. The double vertical lines denote the salt bridge. By convention, the anode, or oxidation half-cell, is written first, to the left of the double lines, and anode, or reduction half-cell, is written to the right of the double lines.

Once you know which electrodes is the anode and which is the cathode, you can determine the direction of electron flow in the external portion of the circuit. Electrons are given up by the anode (from the oxidation half-reaction) and thus flow from it, whereas electrons are used up by the cathode (by the reduction half-reaction) and so flow from it. The anode in a galvanic cell has a negative sign, because electrons flow from it. The cathode in a galvanic cell has a positive sign. In the wire, electrons migrate from the anode to the cathode.

5.6.2 Electrolytic Cells

An **electrolytic cell** is an electrochemical cell in which an external energy source drives an otherwise nonspontaneous reaction. The process of producing a chemical change in an electrolytic cell is called **electrolysis**. Many important substances, including aluminum and chlorine, are produced commercially by electrolysis.

A. Electrolysis of Molten Salts

Electrolysis of molten sodium chloride produces Na and Cl_2 (NaCl melts at 801℃). At the electrode connected to the negative pole of the battery, globules of sodium metal form; chlorine gas evolves from the other electrode. The half-reactions are

$$Na^+(l) + e^- \longrightarrow Na(l)$$

$$Cl^-(l) \longrightarrow \frac{1}{2}Cl_2(g) + e^-$$

As noted earlier, the anode is the electrode at which oxidation occurs, and the cathode is the electrode at which reduction occurs (these definition hold for electrolytic cells as well as for galvanic cells). Thus, during the electrolysis of molten NaCl, the reduction of Na^+ to Na occurs at the cathode, and the oxidation of Cl^- to Cl_2 occurs at the anode.

A **Downs cell** is *a commercial electrochemical cell used to obtain sodium metal by the electrolysis of molten sodium chloride*. The cell is constructed to keep the products of the electrolysis separate, because they would otherwise react. Calcium chloride is added to the sodium chloride to lower the melting point from 801℃ for NaCl to about 580℃ for the mixture (Remember that the melting point, or freezing point, of a substance is lowered by the addition of a solute). You obtain the cell reaction by adding the half-reactions.

$$Na^+(l) + e^- \longrightarrow Na(l)$$

$$Cl^-(l) \longrightarrow \frac{1}{2}Cl_2(g) + e^-$$

$$\overline{Na^+(l) + Cl^-(l) \longrightarrow Na(l) + \frac{1}{2}Cl_2(g)}$$

A number of other reactive metals are obtained by electrolysis of a molten salt or an ionic compound. Lithium, magnesium, and calcium metals are all obtained by the electrolysis. The first presentation of sodium metal adapted the method used by Humphry Davy when discovered the element in 1807. Davy electrolyzed molten sodium hydroxide, NaOH, whose melting point (318℃) is relatively low for an ionic compound. The half-reactions are

$$Na^+(l) + e^- \longrightarrow Na(l) \quad \text{(cathode)}$$
$$4OH^-(l) \longrightarrow O_2(g) + 2H_2O(g) + 4e^- \quad \text{(anode)}$$

B. Electrolysis of Sodium Chloride Solutions

When you electrolyze an aqueous solution of sodium chloride, NaCl, the possible species involved in half-reactions are Na^+, Cl^-, and H_2O. The possible cathode half-reactions are

$$Na^+(aq) + e^- \longrightarrow Na(s) \quad E^\ominus = -2.71V$$
$$2H_2O(l) + 2e^- \longrightarrow H_2(g) + 2OH^-(aq) \quad E^\ominus = -0.83V$$

Under standard conditions, you expect H_2O to be reduced in preference to Na^+, which agrees with what you observe. Hydrogen gas evolves at the cathode.

The possible anode half-reactions are

$$2Cl^-(aq) \longrightarrow Cl_2(g) + 2e^- \quad -E^\ominus = -1.36V$$
$$2H_2O(l) \longrightarrow O_2(g) + 4H^+(aq) + 4e^- \quad -E^\ominus = -1.23V$$

Under standard-state conditions, you expect H_2O to be oxidized in preference to Cl^-. However, the potentials are close and overvoltages at the electrodes could alter this conclusion.

It is possible, nevertheless, to give a general statement about the product expected at the anode. Electrode potentials, as you have seen, depend on concentrations. It turns out that when the solution is concentrated enough in Cl^-, Cl_2 is the product, but in dilute solution, O_2 is the product. To see this, you would simply apply the Nernst equation to the Cl^-/Cl_2 half-reaction.

$$2Cl^-(aq) \longrightarrow Cl_2(g) + 2e^-$$

Starting with very dilute NaCl solution, you would find the oxidation potential of Cl^- is very negative, so H_2O is reduced in preference to Cl^-. But as you increased the NaCl concentration, you would find that the oxidation potential of Cl^- increases until eventually Cl^- is oxidized in preference to H_2O. The product changes from O_2 to Cl_2.

The half-reaction and cell reaction for the electrolysis of aqueous sodium chloride to chlorine and hydroxide ion are as follows:

$$2H_2O(l) + 2e^- \longrightarrow H_2(g) + 2OH^-(aq)$$
$$2Cl^-(aq) \longrightarrow Cl_2(g) + 2e^-$$

$$2H_2O(l) + 2Cl^-(aq) \longrightarrow H_2(g) + Cl_2(g) + 2OH^-(aq)$$

Because the electrolysis started with sodium chloride, the cation in the electrolyte solution is Na^+. When you evaporate the electrolyte solution at the cathode, you obtain sodium hydroxide, NaOH.

The electrolysis of aqueous sodium chloride is the basis of the **chlor-alkali industry**, the major commercial source of chlorine and sodium hydroxide. Commercial cells are of several types, but in each the main problem is to keep the produces separate, because chlorine reacts with aqueous sodium hydroxide.

A modern **chlor-alkali membrane cell** is *a cell for the electrolysis of aqueous sodium chloride in which the anode and cathode compartments are separated by a special plastic* *membrane that allows only cations to pass through it*. Sodium chloride solution is added to the anode compartment, where chloride ion is oxidized to chlorine. The sodium ions carry the current from the anode to the cathode by passing through the membrane. Water is added at the top of the cathode compartment, where it is reduced to hydroxy ion, and sodium hydroxide solution is removed at the bottom of the cathode compartment.

The older chlor-alkali mercury cell is a cell for the electrolysis of aqueous sodium chloride in which mercury metal is used as the cathode. At the mercury cathode, sodium ion is reduced in preference to water. Sodium ion is reduced to sodium to form a liquid sodium-mercury alloy called amalgam (An amalgam is an alloy of mercury with any various other metals).

$$Na^+(aq) + e^- \longrightarrow Na(amalgam)$$

Sodium amalgam circulates from the electrolytic cell to the amalgam decomposer. The amalgam and graphite particles in the decomposer form the electrodes of many galvanic cells, in which sodium reacts with water to give sodium hydroxide solution and hydrogen gas. Historically, loss of mercury from these cells has been a source of mercury pollution of waterways.

C. Electroplating of Metals

Many metals are protected from corrosion by plating them with other metals. Zinc coatings are often used to protect steel, because the coat protects the steel by cathodic protection even when the zinc coat is scratched. A thin zinc coating can be applied to steel by **electrogalvanizing**, or zinc electroplating (Galvanized steel has a thick zinc coating obtained by dipping the object in molten zinc). The steel object is placed in a bath of zinc salts and made the cathode in an electrolytic cell. The cathode half-reaction is

$$Zn^{2+}(aq) + 2e^- \longrightarrow Zn(s)$$

Electrolysis is also used to purify some metals. For examples, copper for electrical use, which must be very pure, is purified by electrolysis. Slabs of impure copper serve as anode, and pure copper sheets serve as cathodes; the electrolyte bath is copper(II) sulfate ($CuSO_4$). During the electrolysis, copper(II) ions leave the anode slabs and plate out on the cathode sheets. Less reactive metals, such as gold, silver, and platinum, that were present in the impure copper form a valuable mud that collects on the bottom of the electrolytic cell. Metals more reactive than copper remain as ions in the electrolytic bath. After about a month in the electrolytic cell, the pure copper cathodes are much enlarged and are removed from the cell bath.

New Words and Expressions

amalgam /əˈmælgəm/ n. 汞合金；汞齐
anode /ˈænəʊd/ n. 负极；阳极
cathode /ˈkæθəʊd/ n. 正极；阴极
cell diagram n. 电池图示
cell potential n. 电池电动势
chlor-alkali industry n. 氯碱工业

Downs cell n. 唐斯电解池
electrochemical cell n. 电化学电池
electrode /ɪˈlektrəʊd/ n. 电极
electrogalvanizing /ɪˈlektrəʊˈgælvənaɪzɪŋ/ n. 电解镀锌
electrolysis /ɪˌlekˈtrɒləsəs/ n. 电解

electrolytic cell n. 电解池
electroplating /ɪˈlektrəʊˌpleɪtɪŋ/ n. 电镀
galvanic cell n. 伽伐尼电池
globule /ˈglɒbjuːl/ n. 小滴；小球体
half-cell /hɑːf sel/ n. 半电池
half-reaction /hɑːf riˈækʃn/ n. 半反应

mud /mʌd/ n. 泥
overvoltage /ˌəʊvəˈvəʊltɪdʒ/ n. 过电势
salt bridge n. 盐桥
slab /slæb/ n. 厚板
voltaic cell n. 伏打电池

Notes

1. 与 electro- 相关的词：electrochemical（电化学的），electrolysis（电解），electrolytic（电解的），electrode（电极），electrolyte（电解质），electroplating（电镀），electron（电子）。

2. anode 及 cathode 在电极反应中分别进行氧化及还原反应。在原电池中，anode 及 cathode 分别翻译为负极及正极，在电解池中，anode 及 cathode 分别翻译为阳极及阴极。

3. An electrochemical cell is a system consisting of electrodes that dip into an electrolyte and in which a chemical reaction either uses or generates an electric current.

参考译文：电化学电池是一种由浸入电解液的电极组成的系统，化学反应在该系统中使用或产生电流。

4. The cell operates on the principle that the oxidation of Zn to Zn^{2+} and the reduction of Cu^{2+} to Cu can be made to take place simultaneously in separate locations with the transfer of electrons between them occurring through an external wire.

参考译文：该电池的工作原理为在不同的位置同时发生锌的氧化（产生锌离子）及铜离子的还原（产生铜），并且经由外部电线进行它们的电子转移。

5. Without the salt bridge connecting the two solutions, the buildup of positive charge in the anode compartment (due to the departure of electrons and the resulting formation of Zn^{2+}) and the buildup of negative charge in the cathode compartment (created by the arrival of electrons and the reduction of Cu^{2+} to Cu) would quickly prevent the cell from operating.

参考译文：如果没有连接这两种溶液的盐桥，正电荷在负极隔间的累积（由于电子离去和由此形成的锌离子）及负电荷在正极隔间的累积（由于电子的到来和铜离子的还原为铜）很快就会阻止电池运行。

6. Electrons are given up by the anode (from the oxidation half-reaction) and thus flow from it, whereas electrons are used up by the cathode (by the reduction half-reaction) and so flow from it.

参考译文：由于负极进行氧化半反应，电子从负极流出，而正极进行还原半反应，电子从正极流入。

7. A modern chloro-alkali membrane cell is a cell for the electrolysis of aqueous sodium chloride in which the anode and cathode compartments are separated by a special plastic membrane that allows only cations to pass through it.

参考译文：现代氯碱膜电解池是一种电解氯化钠溶液的电解池，其中阳极及阴极隔层由一种只允许阳离子通过的特殊塑料膜隔开。

Questions

1. What is an electrochemical cell?
2. What are the components of a galvanic cell?

Chapter 6

Chemistry in Life

6.1 Nitroglycerine

Nitroglycerin (Figure 6.1), a trinitrate of glycerol (propane-1, 2, 3-triol), is an explosive liquid. It is made by treating glycerol with a mixture of concentrated sulfuric acid and nitric acid.

Nitroglycerine was discovered in Italy in 1846 but it was too dangerous in its pure form to be used in practice since it is extremely sensitive to shock. With just a little jostling, nitroglycerine can rearrange its atoms to give stable products

$$4C_3H_5N_3O_9(l) \longrightarrow 6N_2(g) + 12CO_2(g) + O_2(g) + 10H_2O(g)$$

The stability of the products results from their strong bonds, which are much stronger than those in nitroglycerin. Nitrogen, for example, has a strong nitrogen-nitrogen triple bond, and carbon dioxide has two strong carbon-oxygen double bonds. The explosive force of the reaction results from both rapid reaction and from the larger volume increase on forming gaseous products. Among the students working on nitroglycerine was Alfred Nobel, a young Swedish chemist and engineer. When he returned home to Sweden, he carried on his research and set up a family business making explosives.

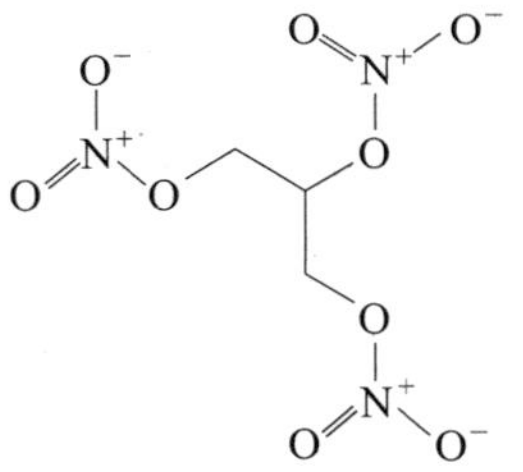

Figure 6.1 Structure of nitroglycerine

Handing nitroglycerine was dangerous and a number of workers, including Alfred's younger brother Emil, were killed. Nobel's research led him to discover that nitroglycerine could be adsorbed onto solids such as kieselguhr, a soft chalk-like rock found in the hills near one of his factories in Germany. Mixing nitroglycerine with kieselguhr, or other solids such as sawdust or with sand or clay, made a useful explosive, **dynamite**. This is safer and more easily handled since it does not explode until detonated.

Dynamite is still one of the most commonly used explosives in the mining and construction industries. The invention made Nobel a very rich man and he founded the Nobel Institute, which to this day awards the prizes that are named in his honor.

In addition to being an explosive, nitroglycerine also serves as a vasodilator (dilates blood vessels) and therefore reduces the chances of blockage that leads to a heart attack. Louis Ignarro, a scientist at UCLA, investigated the action of nitroglycerin in the body. He discovered that metabolism of nitroglycerin produces nitric oxide, NO, which is responsible for a large number of physiological processes. He was awarded the 1998 Nobel Prize in Physiology or Medicine for this discovery.

New Words and Expressions

blood vessels n. 血管
clay /kleɪ/ n. 黏土
detonate /ˈdetəneɪt/ v. （使）爆炸；引爆
dilate /daɪˈleɪt/ v. 膨胀，扩张
dynamite /ˈdaɪnəmaɪt/ n. 黄色炸药
glycerol /ˈɡlɪsərɒl/ n. 甘油
heart attack n. 心脏病发作

jostling /ˈdʒɒs lɪŋ/ v. 推，撞
kieselguhr /ˈkiːzəlˌɡʊr/ n. 硅藻土
nitroglycerin /ˌnaɪtrəʊˈɡlɪsərɪn/ n. 硝化甘油
physiological /ˌfɪziəˈlɒdʒɪkl/ adj. 生理学的
physiology /ˌfɪziˈɒlədʒi/ n. 生理学
sawdust /ˈsɔːdʌst/ n. 锯末
vasodilator /ˌveɪzəʊˈdaɪleɪtə/ n. 血管扩张剂

Questions

1. How can nitroglycerin be made?
2. Which substances were added in nitroglycerine to make it more stable?
3. What are uses of nitroglycerin in our life?

6.2 Coupling of Reactions

Many important processes are thermodynamically unfavorable, or nonspontaneous. One of nonspontaneous processes in biological systems is assembling many small amino acid molecules into a larger protein molecule. Yet biological organisms do assemble proteins from amino acids. Biological organisms accomplish many similar nonspontaneous processes. How do they do this?

Let's look at a simple example. We would like to obtain iron metal from iron ore. Here is the equation for the direct decomposition of iron(Ⅲ) oxide, the iron-containing substance in the iron ore hematite

$$2Fe_2O_3(s) \longrightarrow 4Fe(s) + 3O_2(g) \qquad \Delta G^\ominus = +1487 \text{ kJ/mol}$$

This reaction is clearly nonspontaneous, because $\Delta G^{\ominus}$ is a large positive quantity. The nonspontaneity of this reaction is in agreement with common knowledge. Iron tends to rust (forms iron oxide) in air, by combining with oxygen, so you do not expect a rusty wrench to turn spontaneously into shiny iron and oxygen. But this does not mean you cannot change iron(Ⅲ) oxide to iron metal. It merely means that you have to do work on the iron(Ⅲ) oxide to reduce it to the metal. You must, in effect, find a way to couple this nonspontaneous reaction to one that is sufficiently spontaneous; that is, you must couple this reaction to one having a more negative $\Delta G^{\ominus}$.

Consider the spontaneous reaction of carbon monoxide with oxygen (the burning of CO)
$$2CO(g) + O_2(g) \longrightarrow 2CO_2(g) \qquad \Delta G^{\ominus} = -514.4 \text{ kJ/mol}$$
For 3 mol O_2, $\Delta G^{\ominus}$ is -1543 kJ, which is more negative than for the direct decomposition of 2 mol Fe_2O_3 to its elements. Let us add the two reactions

$$2Fe_2O_3(s) \longrightarrow 4Fe(s) + 3O_2(g) \qquad \Delta G^{\ominus} = +1487 \text{ kJ/mol}$$
$$6CO(g) + 3O_2(g) \longrightarrow 6CO_2(g) \qquad \Delta G^{\ominus} = -1543 \text{ kJ/mol}$$

$$2Fe_2O_3(s) + 6CO(g) \longrightarrow 4Fe(s) + 6CO_2(g) \qquad \Delta G^{\ominus} = -56 \text{ kJ/mol}$$

The net result is to reduce iron(Ⅲ) to the metal by the reaction of iron(Ⅲ) oxide with carbon monoxide. This is the reaction that occurs in a blast furnace, where iron ore is commercially reduced to iron.

The concept of the coupling of two chemical reactions, a nonspontaneous reaction with a spontaneous one, to give an overall spontaneous change is a very useful one, especially in biochemistry. Adenosine triphosphate, or ATP, is a molecule containing a triphosphate group. It reacts with water, in the presence of an enzyme (biochemical catalyst), to give adenosine diphosphate, ADP, and a dihydrogen phosphate ion. Figure 6.2 shows the structures of ATP and ADP. The reaction has a negative $\Delta G^{\ominus}$ and it is a spontaneous reaction.
$$ATP + H_2O \longrightarrow ADP + H_2PO_4^- \qquad \Delta G^{\ominus} = -31 \text{ kJ/mol}$$

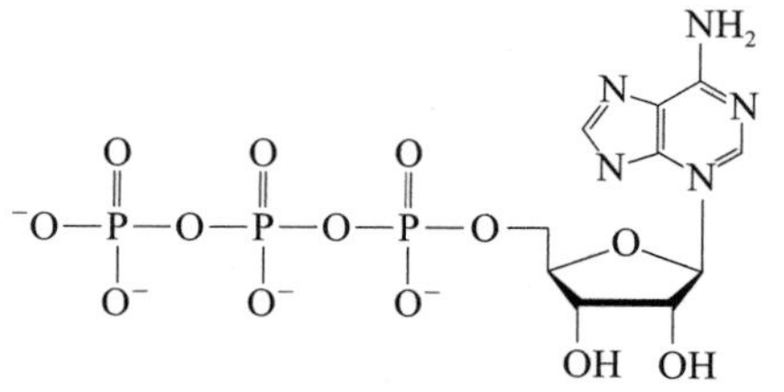

(a) ATP 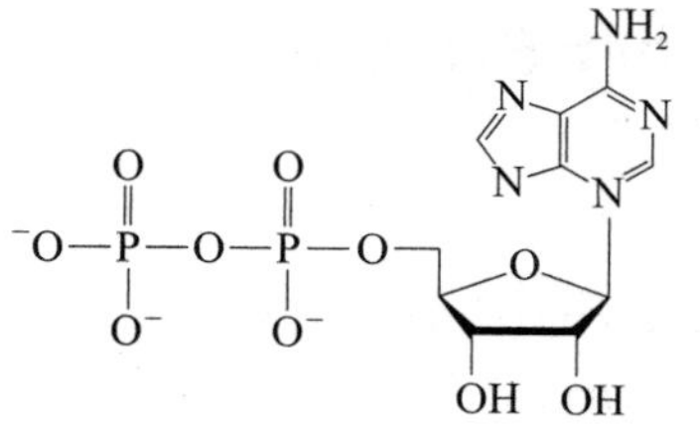 (b) ADP

Figure 6.2 Structures of ATP and ADP

ATP is first synthesized in a living organism, and the energy for this synthesis is obtained from food (oxidation of glucose). The spontaneous reaction of ATP with water to give ADP and dihydrogen phosphate ion is then coupled to various nonspontaneous reactions in the organism to accomplish necessary life processes. For example, the stepwise synthesis of a protein molecule involves the joining of individual amino acids. Let's consider the formation of the dipeptide (a unit composed of two amino acids) alanylglycine from alanine and glycine.

The reaction represents the first step in the synthesis of a protein molecule.

$$\text{alanine} + \text{glycine} \longrightarrow \text{alanylglycine} \quad \Delta G^\ominus = 29 \text{ kJ/mol}$$

The positive $\Delta G^\ominus$ value means the production of alanylglycine is not favorable. However, with the aid of an enzyme, the reaction is coupled to the hydrolysis of ATP as follows

$$\text{ATP} + \text{H}_2\text{O} + \text{alanine} + \text{glycine} \longrightarrow \text{ADP} + \text{H}_2\text{PO}_4^- + \text{alanylglycine}$$

The overall free energy change for this coupled reaction is $\Delta G^\ominus = -2$ kJ/mol. Thus, the coupled reaction now favors the production of alanylglycine.

New Words and Expressions

adenosine diphosphate n. 二磷酸腺苷
adenosine triphosphate n. 三磷酸腺苷
alanine /ˈæləniːn/ n. 丙氨酸
amino acid n. 氨基酸
blast furnace n. 鼓风炉
decomposition /ˌdiːkɒmpəˈzɪʃn/ n. 分解
enzyme /ˈenzaɪm/ n. 酶

glycine /ˈɡlaɪsiːn/ n. 甘氨酸
hematite /ˈhiːmətaɪt/ n. 赤铁矿
ore /ɔː(r)/ n. 矿石
protein /ˈprəʊtiːn/ n. 蛋白质
rusty /ˈrʌsti/ adj. 生锈的
spontaneous /spɒnˈteɪniəs/ adj. 自发的
wrench /rentʃ/ n. 扳钳

Questions

1. How can a nonspontaneous reactions become a spontaneous reaction? Give an example.

2. Why can ATP act as "energy currency" in the synthesis of proteins from amino acids?

6.3 Haber Process

Nitrogen-containing compounds are very useful chemicals. For example, they can be used as agricultural fertilizers and as explosives. You might think there is little to link these applications-but in fact there is a common factor that provides an interesting example of how chemical equilibria are considered in industrial processes.

Until the early part of the twentieth century, the main sources of nitrogen compounds for use as fertilizers were sodium nitrate and potassium nitrate, which were mined from deposits in the ground. In the early twentieth century, the food requirements of an increasing population in Europe meant that mining could no longer provide sufficient nitrates for agricultural needs. In 1912, the German chemist Fritz Haber developed a process to "fix" atmospheric nitrogen to manufacture ammonia, NH_3

$$N_2(g) + 3H_2(g) \rightleftharpoons 2NH_3(g)$$

Since ammonia can be converted by soil bacteria to nitrates and nitrates that can be used by plants, one of the principal uses of ammonia is as a fertilizer that is injected directly into the soil. Ammonia is also used in the production of other nitrogen compounds, such as ammonium nitrate, ammonium hydrogen phosphate, ammonium dihydrogen phosphate, ammonium sulfate, and hydrazine. Several of these compounds are used as fertilizers, and others are used in the manufacture of explosives, pharmaceuticals, and plastics.

Haber realized that, in order to maximize the production of ammonia, the reaction should be operated at high pressure (Le Chaterlier's principle). In addition, the removal of ammonia as it is formed favors the forward reaction. The forward reaction is exothermic and so, in principle, should be more efficient at low temperatures. However, at low temperatures, the reaction is too slow to be used commercially and an important discovery for the production of large amounts of ammonia was an iron-based catalyst that allows the reaction to proceed at usable rates. Haber used a kinetic and thermodynamic analysis of the reactions to determine that the optimum conditions were 450℃ and 200 atm (1 atm = 101325 Pa) pressure. The process is sometimes known as the "Haber-Bosch" process to recognize the contribution of Karl Bosch, the engineer who designed and built the plant to operate under these (in those days) challenging conditions. The ammonia produced can be converted to other useful compounds for fertilizers and explosives. Today, most reactors for ammonia synthesis operate between 25 atm and 150 atm.

Shortly after its discovery the Haber process became even more vital to Germany. At the outbreak of the First World War in 1914, Germany relied on imports of nitrates from South America. These were cut off by a naval blockade so the Haber process became the source of all nitrogen compounds for fertilizers and also for the explosives in weapons used by the military. Haber was awarded the 1918 Nobel Prize in Chemistry for his discovery of the catalysts and optimization of the nitrogen fixation reaction. This was controversial, however, since he was also instrumental in developing the manufacture of poisonous gases such as chlorine and phosgene, used as weapons during the war.

The Haber process remains the major source of ammonia to this day and is of tremendous commercial significance, with a worldwide production in 2017 estimated as 2.9×10^{11} kg.

New Words and Expressions

blockade /blɒˈkeɪd/ n. 封锁
controversial /ˌkɒntrəˈvɜːʃl/ adj. 有争议的
deposit /dɪˈpɒzɪt/ n. 沉积物
exothermic /ˌeksəʊˈθɜːmɪk/ adj. 放热的
explosive /ɪkˈspləʊsɪv/ n. 炸药
forward reaction n. 正向反应

hydrazine /ˈhaɪdrəˌziːn/ n. 肼
mine /maɪn/ v. 采矿
naval /ˈneɪvl/ adj. 海军的
pharmaceutical /ˌfɑːməˈsuːtɪkl/ n. 药物
phosgene /ˈfɒzdʒiːn/ n. 光气
plastic /ˈplæstɪk/ n. 塑料

> **Questions**

1. Why is ammonia an important source for fertilizers?
2. How does Haber increase the production yield of ammonia?

6.4 Air Bags

An air bag, a supplementary restraint system (SRS) in automobiles, has saved thousands of lives in car accidents. The idea behind this is simple. When a crash occurs, a plastic bag rapidly inflates with a gas, protecting the driver from hitting the dashboard or steering column. However, the development of a workable air-bag system required the combined efforts of chemists and engineers.

There are some special requirements in air-bag system. First, the air bag must not inflate accidentally. Second, the gas used must be nontoxic in case of leakage after inflation. Third, the gas must be "cool", so as not to produce burn injuries. Fourth, the gas must be produced very rapidly, ideally inflating the bag within 20~60 ms. Finally, the gas-producing chemicals must be easy to handle and be stable for long periods.

Nitrogen is the best choice for nontoxic gas, and it exists in the atmosphere we live in. There are several gas generation systems in use, but a common one uses the deposition of alkali metal azidse, such as sodium azide, NaN_3.

$$2NaN_3(s) \xrightarrow{\triangle} 2Na(l) + 3N_2(g)$$

Sensors that detect the initial crash activate the air-bag system by electrically initiating the explosion of a small charge. This, in turn, starts the rapid burning of a pellet containing sodium azide, which releases a large volume of $N_2(g)$ to fill the bag.

By 1980, the engineering challenges of the air-bag safety system had been resolved, but not the chemical challenges. The problems posed by the use of sodium azide were that it does not make a good pellet, its reaction does not go to completion rapidly, and one of the reaction products—sodium metal—reacts violently with water.

To solve these problems, investigators tried adding other compounds to the sodium azide. To make a good pellet-forming mixture, a lubricant, typically molybdenum disulfide (MoS_2), was added to the sodium azide. This mixture did not burn well, however. Sulfur, a familiar constituent of gunpowder, was then added to produce smooth-burning pellets. The nitrogen gas produced was cool, and the sodium metal was converted mainly to the sulfate, though the solid residue was finely powdered and difficult to contain. Some air-bag systems on the market today use the MoS_2-S-NaN_3 pellet, but the latest ones use a pellet that is still more complex.

In earlier experimental work, a mixture of sodium azide and iron(Ⅲ) oxide had proved satisfactory for trapping the sodium metal by converting it to an easy-to-handle solid residue. This mixture did not burn well, however. Researcher then tried the obvious solution: Mix together all the compounds that give the gas-forming pellets their most desirable properties: sodium azide, iron(Ⅲ) oxide, molybdenum disulfide, and sulfur. In scientific research, the obvious often turns up some unexpected results. In this case, however, the hoped-for final result was achieved: a rapidly burning pellet producing cool, order-free nitrogen gas and a nonreactive solid residue that is easily trapped. At this point, the widespread use of air-bag collision systems in automobiles became a reality.

New Words and Expressions

dashboard /ˈdæʃbɔːd/ n. 仪表板
inflate /ɪnˈfleɪt/ v. 膨胀
lubricant /ˈluːbrɪkənt/ n. 润滑剂
molybdenum disulfide n. 二硫化钼
pellet /ˈpelɪt/ n. 颗粒状物；小球

residue /ˈrezɪdjuː/ n. 残留物
steering column n. 转向柱
supplementary restraint system n. 辅助约束系统

Questions

1. How can air bags save lives in car accidents?
2. What is the gas produced in the air bag? What is the chemical equation for producing this gas?
3. What is the problem when using sodium azide in the air bag?

6.5 Self-Cleaning Windows

Pilkington Activ™, launched in 2001, was the first type of glass to be produced that has the ability to clean itself. The glass contains a virtually transparent coating of titanium dioxide, 15 nm thick, which is deposited during the manufacturing process. The TiO_2 coating is durable, since it is bonded to the glass surface, and has two functions that allow it to act as a self-cleaning glass.

Firstly, extremely fine particles of TiO_2 on the glass absorbs ultraviolet photons from sunlight. On absorption of a photo, the electron in TiO_2 is promoted from the filled valence band into the empty conduction band. The promoted electron, which is in the higher energy level, is then able to interact with oxygen adsorbed on the surface to produce a superoxide ion, O_2^-.

$$O_2 + e^- \longrightarrow O_2^-$$

Once titanium dioxide has been activated in this way, the photoactive form of titanium dioxide acts as an oxidizing agent by accepting electrons into vacancies in the valence band. It obtains these electrons by oxidation of water, converting H_2O into very reactive hydroxyl radicals, $OH·$.

$$H_2O \longrightarrow OH· + H^+ + e^-$$

The hydroxyl radicals and superoxide ions are both strong oxidizing agents and are able to oxidize most of the organic molecules present in dirt, converting them eventually to carbon dioxide and water. This means that photoactivation of TiO_2 provides a means for getting rid of most organic dirt.

The second way in which the TiO_2 coating leads to self-cleaning relates to interactions with water molecules. On the surface of the coating, the oxygen atoms of TiO_2 are protonated, forming OH groups. These groups are hydrophilic, and interact with water molecules through hydrogen bonds. As a result, rain water spreads out into a thin film on the glass surface, and dirt is washed off the window in a sheet of water.

On the contrary, normal glass is hydrophobic. Water forms droplets on most types of glass, and these run off the surface in streams. The streams tend to concentrate the dirt, and lead to smudges and drying masks.

Not only glass can benefit from the free radical generating capability of titanium dioxide, but also cement can be kept from graying. The Jubilee Church in Rome, completed in 2003, is one of the examples of this technology. Made of self-cleaning concrete, it is designed to last for a thousand years, hopefully maintaining its white luster. Air-borne substances can be decomposed as well when they contact a light activated titanium dioxide surface.

New Words and Expressions

cement /sɪˈment/ n. 水泥
conduction band n. 导带
free radical /ˌfriːˈrædɪk(ə)l/ n. 自由基
hydrophilic /ˌhaɪdrəˈfɪlɪk/ adj. 亲水的
hydrophobic /ˌhaɪdrəˈfəʊbɪk/ adj. 疏水的
valence band n. 价带

Questions

1. What is the principle for the self-cleaning window?
2. What is the difference between the normal glass and the self-cleaning window?

6.6 Lithium-Ion Batteries

Ionic size plays a major role in determining the properties of devices that rely on movement of ions. For example, lithium-ion batteries are used in many devices, such as cell phones, iPods, and laptop computers. How do these lithium-ion batteries work? What are their advantages for these portable devices?

Like similar electrochemical cells, lithium-ion cells have an anode and a cathode separated by an electrolyte. But the detailed structure for lithium-ion cells is rather different from that of a simple galvanic cell, in which the electrodes are metals. Each electrode of a lithium-ion cell is composed of a material having a layer structure, one layer on top of another, between which lithium ions can be reversibly inserted or extracted.

The actual electrode and electrolyte materials depend on the particular battery, but a common lithium-ion cell has one electrode composed of lithium cobalt oxide, $LiCoO_2$, and the other electrode of graphite, a form of carbon. Lithium cobalt oxide has a layer structure, with layers of lithium ions, Li^+, alternating with layers of cobalt oxide, CoO_2^-. Graphite has a layer structure in which one sheet of covalently bonded carbon atoms is normally held to another sheet by van der Waals forces. During battery charging, lithium ions move from the $LiCoO_2$ electrode to the graphite electrode, lodging between graphite sheets, while electrons from the external circuit give some of the carbon atoms a negative charge to balance that of the lithium ions. Ionic bonds form between the lithium ions and negatively charged carbon atoms. The half-reactions during charging, in which a fraction of the lithium ions, x, moves from the $LiCoO_2$ electrode to the graphite electrode, are

$$LiCoO_2(s) \longrightarrow Li_{1-x}CoO_2(s) + xLi^+ + xe^-$$
$$C(s) + xLi^+ + xe^- \longrightarrow Li_xC(s)$$

These half-reactions are reversed during discharge, so that the graphite electrode is the anode (becoming the negative electrode) and the lithium cobalt oxide electrode is the cathode (the positive electrode).

The electrolyte is a nonaqueous solution of a lithium salt, for example, lithium hexafluorophosphate, $LiPF_6$. (A nonaqueous solvent is required, because the voltage in a lithium-ion cell is higher than what would electrolyze water.) The electrolyte is partitioned by a microporous polymer sheet, or separator, to keep the electrodes apart and allow lithium ions, but not electrons, to pass through.

The environment of the lithium ions at the two electrodes are different, so the free energies of the ions at the anode and cathode are different. During discharge, lithium ions flow from the graphite electrode, where the ions have high free energy, through the electrolyte to the $LiCoO_2$ electrode, where the ions have lower free energy. This decrease of free energy of the lithium ions is the source of the battery power.

Lithium is a very light element, and it is also very reactive, so it can store more energy for a given mass and size. High energy storage for given mass and size of battery is a special requirement of portable devices. Cell phones, tablets, and laptop computers all use lithium-ion battery for this reason. Gas-electric hybrid cars and plug-in electric cars are another use for these batteries. To be competitive with fossil fuels, scientists are trying to discover new cathode and anode materials that will easily accept and release lithium ions without falling apart over many repeated cycles. In addition, new separator materials that allow for faster lithium ion passage are also under development. Some research groups are looking at using sodium ions instead of lithium ions because sodium is far more abundant on Earth than lithium; new materials that allow sodium ion insertion and release are therefore under development.

New Words and Expressions

discharge /dɪsˈtʃɑːdʒ/ v. 放电
graphite /ˈɡræfaɪt/ n. 石墨
microporous /ˈmaɪkrouˈpɔːrəs/ adj. 微孔性的
van der Waals force n. 范德瓦耳斯力

Questions

1. What are the anode and cathode materials (during discharge) for a common lithium-ion battery, respectively?

2. What are lithium-ion batteries especially desirable for portable devices like cell phones?

6.7 Liquid Crystals

Liquid crystals, as implied by this name, are substances that combine the property of liquids, the ability to flow and take on the shape of a container, with the property of crystals, a regular arrangement of particles in a lattice. Some substances exhibit liquid crystal behavior when they are melted from a solid. When the temperature increases, the solid liquefies but retains some order in one or two dimensions. At a higher temperature, the ordered liquid becomes a more conventional liquid in which there is no consistent orientation of the molecules.

Liquids are **isotropic**, meaning that their properties are independent of the direction of testing. Because particles in a liquid are free to rotate, there are molecules present in every possible orientation, so the bulk sample gives the same measurements regardless of the direction in which the measurements are taken or observations are made. On the contrary, liquid crystals are **anisotropic**, which means that the properties they display depend on the direction (orientation) of the measurement. A box full of pencils, all nearly aligned, is an analogy for anisotropy, because the pencils are long and narrow. Looking at the pencils end on it different than looking at them from the side. What you see depends on the direction from which you view the pencils. Similarly, what you see when you look at anisotropic molecules depends on the direction from which you view them.

Frederick Reinitzer, an Austrian botanist, discovered the first compound to exhibit liquid crystal behavior in 1888. Cholesteryl benzoate was observed to form an oriented liquid crystalline phase when melted, and this liquid crystalline phase became an ordinary liquid at higher temperatures.

cholesteryl benzoate

The structure of cholesteryl benzoate is fairly rigid owing to the presence of the fused rings and sp^2-hybridized carbon atoms. The molecule is relative long compared to its width. Rigidity and this particular shape make it possible for the cholesteryl benzoate molecule to arrange themselves in an orderly manner, in much the way that pencils, chopsticks, or tongue depressors can be arranged.

There are three types of ordering (nematic, smectic, and cholesteric) that a liquid crystal can adapt. In the **nematic** (meaning thread-like) form of the liquid crystalline state, the rod-like molecules are arranged in a parallel fashion. They are free to move in all directions, but they can rotate only on their long axes. (Imagine the ways in which you might move a particular pencil in a box of loosely packed pencils.) In the **smectic** (meaning grease-like) form, rod-like molecules are arranged in layers, with the long axes of the molecules perpendicular to the planes of the layers. The molecular motions possible here are a translation within, but not between, layers and a rotation about the long axis. The **cholesteric** form is related to the smectic form, but the orientation in each layer is different from that in the layer above and below.

Since the commercialization of liquid crystals in the 1970s, many uses for liquid crystals have been developed. Perhaps the best known is in liquid crystal displays (LCDs), which can be used in calculators, watches, and laptop computers, because polarized light can be transmitted through liquid crystals in one phase but not transmitted through another. Polarized light is produced by passing ordinary light (which oscillates in all directions perpendicular to the beam) through a filter that allows only the light waves oscillating in one particular plane to pass through. This plane-polarized light can be rotated by a twisted liquid crystal such that it is allowed to pass through another filter when a voltage is applied to the liquid. If the light pass through the liquid when the voltage is applied to the liquid. If the light pass through the liquid when the voltage is off, though, the light will not pass through. This principle can be used in calculator and watch displays because each digit is made up of no more than the seven segments, and the specific digit depends on which segments are bright and which are dark. LCDs in laptop computers and televisions are more sophisticated, but still operate on the same basic principle.

Liquid crystals can also be used in thermometers. The spacing between crystal layers depends on temperature, and the wavelength of light reflected by the crystal depends on this spacing. Thus, the color can be used to indicate the temperature to which the liquid crystal is exposed.

Liquid crystals occur widely in living matter. Cell membranes and certain tissues have structures that can be described as liquid crystalline. Hardening of the arteries is caused by the deposition of liquid crystalline compounds of cholesterol. Liquid crystalline properties have also been identified in various synthetic polymers, such as Kevlar fiber.

New Words and Expressions

anisotropic /ˌænˌaɪsəʊˈtrɒpɪk/ adj. 各向异性的
cholesteric /ˌkɒlɪˈsterɪk/ n. 胆甾相
cholesterol /kəˈlestərɒl/ n. 胆固醇
cholesteryl benzoate n. 胆甾醇苯甲酸酯
chopstick /ˈtʃɒpstɪk/ n. 筷子
crystalline /ˈkrɪstəlaɪn/ adj. 结晶的
dimension /daɪˈmenʃn/ n. 维
grease-like /ˈɡriːsˌlaɪk/ adj. 油脂的
isotropic /ˌaɪsəʊˈtrɒpɪk/ adj. 各向同性的

Kevlar fiber n. 克维拉纤维
liquid crystal n. 液晶
nematic /nɪˈmætɪk/ n. 向列相
orientation /ˌɔːriənˈteɪʃn/ n. 方向
plane-polarized light n. 平面偏振光
rod-like /ˈrɒdˌlaɪk/ adj. 棒状的
smectic /ˈsmektɪk/ n. 近晶相
thread-like /ˈθredˌlaɪk/ adj. 丝状的；细长的
tongue depressor n. 压舌板

Questions

1. What are liquid crystals?
2. What are the three liquid crystalline forms?
3. What are the application for liquid crystals?
4. How are liquid crystals related to living organisms and human disease?

6.8 Superconductors

No ordinary electrical conductor is perfect. Even metal have some resistance to the flow of electrons, which wastes energy in the form of heat. **Superconductors** have no resistance to the flow of electrons and thus could be very useful for the transmission of electricity over the long distances between power plants and cities and towns. In 1911, the Dutch physicist Heike Kamerlingh Onnes found that mercury lost all its electrical resistance at 4 K, the boiling point of liquid helium. He was awarded the 1913 Nobel prize in Physics for this surprising discovery. However, liquid helium is very expensive, so there are few feasible applications of a mercury superconductor.

Since that time, more superconductors have been discovered, including a lanthanum-, barium-, copper-, and oxygen-containing ceramic compound, and a series of copper oxides ("cuprates"), all of which become superconducting below 77 K, the boiling point of liquid nitrogen (a much more common and less expensive refrigerant than liquid helium). The temperature below which an element, compound, or material becomes superconducting is called the **superconducting transition temperature** (T_c). The higher the T_c, the more useful the superconductor.

In 1986, a new class of high temperature superconductors was discovered. These ceramic materials include $YBa_2Cu_3O_7$, which has a critical temperature of 93 K ($-180℃$). Unlike the superconducting metals, this material is superconducting in liquid nitrogen, which has a boiling point of 77 K ($-196℃$) and is much cheaper than liquid helium. Although 93 K is still cold, it qualifies as quite warm in the world of superconductors. Thus, $YBa_2Cu_3O_7$ is considered to be the first **high-temperature superconductor**. It was first prepared by combining the three metal carbonates at high temperature (1000~1300 K). The oxidation states of copper ions are the mixture of $+2$ and $+3$.

$$4BaCO_3 + Y_2(CO_3)_3 + 4CuCO_3 + Cu_2(CO_3)_3 \longrightarrow 2YBa_2Cu_3O_7 + 14CO_2$$

$YBa_2Cu_3O_7$ has a complex layer structure with CuO_2 sheets separated by "charge reservoirs", and the mechanism of the superconductivity in $YBa_2Cu_3O_7$ and related compounds is still not fully understood.

Like other ceramics, $YBa_2Cu_3O_7$ is brittle. Thus, it is difficult to form into wires. Recently superconducting power cables based on a Bi-Sr-Ca-Cu-O ceramic have been developed and shown to transmit three times the electricity of conventional copper cables.

Another important property of superconductors is their capacity to exclude a magnetic field below their critical temperature. These materials are so strongly diamagnetic that they move away from (or repel) a magnetic field. This means that a superconductor can be used to levitate a magnet or vise versa. This is known as the **Meissner effect** after its discoverer Walter Meissner.

Magnetic levitation is being researched for use in trains, because the amount of friction between the train and the tracks would be drastically reduced in a magnetically levitated ("maglev") train versus one running on ordinary tracks and wheels. The engineering challenge to a maglev to occur, and the highest temperature obtained for superconducting substances are approximately 140 K—warm on the superconductivity scale but still very cold by ordinary standards.

Superconductivity in metals can be explained satisfactorily by BCS theory, first proposed by John Bardeen, Leon Neil Cooper, and John Robert Schrieffer in 1957. They received the Nobel Prize in Physics in 1972 for their work. BCS theory treats superconductivity using quantum mechanical effects, proposing that electrons with opposite spin can pair due to fundamental attractive forces between interference from other atoms and experience no resistance to flow. Superconductivity in ceramics has yet to be satisfactorily explained.

New Words and Expressions

ceramic /sə'ræmɪk/ n. 陶瓷
Meissner effect n. 迈斯纳效应
maglev /'mæglev/ n. 磁悬浮
superconductivity /ˌsuːpəkɒndʌk'tɪvəti/ n. 超导电性

Questions

1. Why is preparing a material which is superconducting at room temperature important?
2. State the applications of superconductors.

6.9 The Discovery of Prodrugs

Azo dyes are currently used in paints, cosmetics, and food. They are also used as dyes for certain bacteria to be seen in the microscopy. In order to find whether azo dyes are toxic to bacteria or not, Fritz Mietzsch and Joseph Klarer began cataloguing azo dyes for possible antibacterial properties. Later, Gerhard Domagk evaluated the dyes for potential activity, which led to the discovery of the potent antibacterial properties of prontosil.

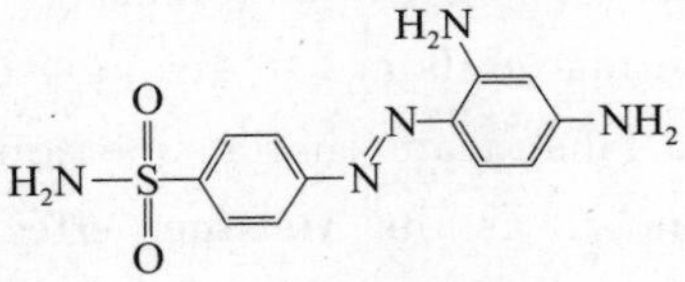

prontosil 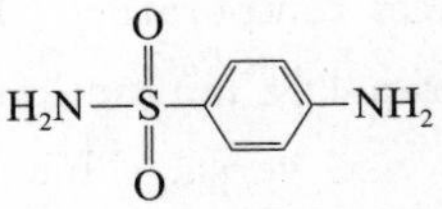   sulfanilamide

Domagk found that prontosil cured streptococcal infections in mice. In 1933, physicians began using prontosil in human patients suffering from life-threating bacterial infections. The success of this drug was extraordinary, and prontosil staked its claim as the first drug that was systematically used for the treatment of bacterial infections. The development of prontosil has been credited with saving thousands of lives. For this pioneering work that led to this discovery, Domagk was award Nobel Prize in Physiology or Medicine.

Prontosil exhibited one very curious property that intrigued scientists. Specifically, it was found to be totally inactive against bacteria *in vitro* (literally "in glass", in bacterial cultures grown in glass dishes). Its antibacterial properties were only observed *in vivo* (literally "in life", when administered to living creatures, such as mice and humans). These observations inspired much research on the activity of prontosil, and in 1935, it was found that prontosil is metabolized in the body to produce a compound called sulfanilamide.

Sulfanilamide was determined to be the active drugs, as it interferes with bacterial cell growth. In a glass dish, prontosil is not converted into sulfanilamide, explaining why the antibacterial properties were only observed *in vivo*. This discovery ushered in the era of prodrugs. **Prodrugs** are pharmacologically inactive compounds that are converted by the body into active compounds. This discovery led scientists to direct their research in new directions. They began designing new potential drugs based on structural modifications to sulfanilamide rather than prontosil. Extensive research was directed at making these sulfanilamide analogues, called sulfonamides. By 1948, over 5000 sulfonamides were created, of which more than 20 were ultimately used in clinical practice.

The emergence of bacterial strains resistant to sulfanilamide, together with the advent of penicillin, rendered most sulfonamides obsolete. Some sulfonamides are still used today to treat specific bacterial infection in patients with AIDS, as well as a few other applications. Despite their small role in current practice, sulfonamides occupy a unique role in history, because their development was based on the discovery of the first know prodrug.

There are currently a large number of prodrugs on the market. Prodrugs are often designed intentionally for a specific purpose. One such drug is used in the treatment of Parkinson's disease. The symptoms of Parkinson's disease are attributed to low level of dopamine in a specific part of the brain. Administering dopamine to a patient does not raise the concentration of dopamine in the brain, because the compound does not readily cross the blood-brain barrier. Instead, the prodrug L-dopa is used because it readily crosses the blood-brain barrier, after which it undergoes decarboxylation to produce the needed dopamine.

There are many different varieties and classes of prodrugs. The discovery of prodrugs was an extremely important achievement in the development of medicinal chemistry, and it all started with a careful analysis of azo dyes.

New Words and Expressions

advent /ˈædvent/ n. 出现
AIDS (Acquired Immune Deficiency Syndrome) 获得性免疫缺乏综合征；艾滋病
azo /ˈæzəʊ/ adj. 偶氮基的
bacterial strain n. 菌株
decarboxylation /ˌdiːkɑːˌbɒksɪˈleɪʃən/ n. 去碳酸基
dopamine /ˈdəʊpəmiːn/ n. 多巴胺
in vitro adj. 在生物体外进行的
in vivo adj. 在生物体内进行的

microscopy /maɪˈkrɒskəpi/ n. 显微术；显微镜观察
obsolete /ˈɒbsəliːt/ adj. 废弃的
Parkinson's disease n. 帕金森病；震颤麻痹
penicillin /ˌpenɪˈsɪlɪn/ n. 盘尼西林（青霉素）
prodrug /ˈprəʊˌdrʌg/ n. 前体药物
prontosil /ˈprɒntəsɪl/ n. 百浪多息（一种偶氮磺胺类药物）
render /ˈrendə(r)/ v. 使得

streptococcal /ˌstreptəˈkɑːkəl/ adj. 链状球菌的

sulfanilamide /ˌsʌlfəˈnɪləmaɪd/ n. 磺胺

sulfonamide /sʌlˈfɒnəmaɪd/ n. 磺酰胺

symptom /ˈsɪmptəm/ n. 症状

usher /ˈʌʃə(r)/ v. 引导

Questions

1. What is the use of prontosil?
2. What is a prodrug?
3. Why is prodrug L-dopa used to treat of Parkinson's disease?

6.10 Magnetic Resonance Imaging

The technique called **magnetic resonance imaging** (MRI) extends the uses of NMR (nuclear magnetic resonance) spectroscopy to investigating complex assemblies of molecules—to the extent that a whole human body can be imagined. Like NMR, MRI depends on nuclear spin process.

MRI was invented to aid medical diagnosis and take advantage of magnetically active hydrogen nucleus which is present in water in aqueous body fluids and fatty tissues. The technique has had such a profound influence on the modern practice of medicine that Paul Lauterbur (a chemist) and Peter Mansfield (a physicist) were awarded the 2003 Nobel Prize in Physiology or Medicine for their discoveries concerning MRI.

This technique relies on using a magnetic field gradient. The protons in two water molecules in the same environment would resonance at the same frequency if placed in the same magnetic field. The energy needed to "flip" the spins and the resonance frequency depend on the strength of the external magnetic field. An MRI scanner, therefore, places the patient in a magnetic field that varies linearly between the poles of the magnet. Two identical protons will now resonance at different frequencies depending on their position in the magnetic field, that is, on their position in the body. A proton in a region where the magnetic field is higher resonates at a higher frequency than in a region where the field is lower.

A particular position in the body is irradiated with a pulse at a particular radiofrequency and the intensity of the emitted radiation is measured. This makes it possible to measure the number of hydrogen nuclei in water at that position. This is repeated at the different positions in the scanner with a succession of pulses at different frequencies to measure the number of hydrogen nuclei at each position. In this way, a "map" of the hydrogen nuclei throughout the body can be built up.

The usual arrangement in a medical scanner is to use a system where the magnetic field has gradients in two dimensions to build up an imagine of a thin "slice" through the patient. A number of such slices can be combined to give a three-dimensional imagine. A "picture" of the body can be then built up with a resolution better than 1 mm. This technique is known as **tomography**.

MRI is a non-invasive diagnostic technique that allows cross-sectional imagines of a human body to be built up without using potentially damaging ionizing radiation such as X-ray. There are differences in water content among tissues and organs and many diseases results in changes of water content. A wide range of diagnoses can be made using MRI including detecting soft-tissue damage and tumors. By looking at how the MRI changes with time, blood flow can be monitored.

In addition to medical applications, MRI has also been applied to a number of chemical problems-for example, in mapping the reactive sites of catalysts and investigating mixing and flow patterns in fluid mixtures.

New Words and Expressions

fatty tissue n. 脂肪组织
irradiate /ɪˈreɪdieɪt/ v. 照射
magnetic resonance imaging n. 核磁共振成像
tomography /təˈmɒɡrəfi/ n. 断层扫描

Questions

Why are hydrogen nuclei, rather than other NMR active nuclei, chosen for monitoring the MRI technique?

Chapter 7
Writing Lab Reports and Papers

7.1 Some Guidelines for Writing Lab Reports

Scientific communications, including lab reports generally contain the following sections that appear in order: abstract, introduction (or principle), methods (or experimental), results (or data and calculations), discussion, conclusion and appendix (if necessary). The introduction, methods, data and analysis, and discussion and conclusion sections make up the body of the document. The abstract stands alone because it simply summarizes the other sections. Below is a detailed description of what should be written in each of the sections. The experimental report is illustrated as follows.

> **Title:** The title of the experiment
> **Date:** month/day/year
> **Author:** name
> **Partner:** name
> **Section:**
> **Abstract (about 2 lines)**
>   The abstract is a bit like a movie preview or the back cover of a book in that it is designed to help the reader to make a judgment about whether the rest of the paper is worth reading. It summarizes each of the sections. Therefore, it should always be the last section that is written, even though it appears in the document first. A simple recipe for writing an effective abstract is to start with a sentence that summarizes the experiment that was performed. What has been done in the lab? It is a brief summary of the results and conclusions.
> **1. Introduction (about 1 paragraph)**
>   In a lab report the nature of the introduction section is a little bit different than in a

research article. In a lab report, you should focus the introduction on the learning goals of the experiment. Discuss how the experiment is designed to achieve these learning goals and how the experiment fits in with the broader curriculum of the corresponding lecture course.

2. Methods (Experimental) (about 1 paragraph)

This section should provide the details of how the experiment was carried out. It should not be written as a recipe but more as a journal entry; a fairly detailed account of what was done in lab. A description of how the data was processed should also be part of the methods section.

3. Results (Data and Calculations)

Whenever appropriate your data should be displayed in tables and figures. The figures and tables should have captions that describe what they are illustrating. You should also prepare sentences that introduce the tables and figures and describe what they show. To an extent these sentences and the captions will be and should be somewhat redundant.

4. Discussion (about one page)

The discussion section will discuss the significance of the findings from the data analysis section. It is also in this section the questions that are being asked are addressed in the context of a well-written paragraph. The following items may be included in this section:

(ⅰ) The conclusion in terms of final calculated values.

(ⅱ) Interpretation of the final results.

(ⅲ) Error propagation.

(ⅳ) Limitation of the experimental method and suggestion of possible improvements.

5. Conclusion

This section should consist of at least two paragraphs that summarize what happened in the experiment, what you learned while doing the experiment, and what your data and results mean. Please remember that your interpretation of the experiment and its results is the most important part of your lab report.

6. Reference

A list of relevant papers, handbooks or websites can be included in this section.

7. Appendix (if necessary)

Spectrum and long calculations can be appended at the end of the lab report.

Notes on writing style: One of the most challenging things for many students is learning the art of writing in the passive voice. Lab report writing is a form of technical writing. It is not like other works you are used to producing, such as email messaging, letter writing or even English composition papers. In science, the experimenter is immaterial. The experimental details and results are what is important. So, never use any pronouns. Also, do not waste words. For instance, In this lab we⋯or the goal of this experiment was⋯or we determined the… can be omitted. Good science writing is written in relative short, clear and concise sentences. It should read a lot more like a newspaper article than a novel.

An example of laboratory datasheet and lab report were shown in the next few pages to illustrate what we should prepare for the experiment (prelab datasheet), what we should record during the experiment (experiment datasheet) and what we should write for a lab report.

Prelab datasheet

Course: CHM192 Organic Experiment (Ⅰ)   Date: 04 / 28 / 2021
Name: Wang Xiaolun
Title: Isolation of Eugenol by Steam Distillation of Cloves
Purpose: To isolate eugenol from cloves by steam distillation and analyze it using mass, ^{13}C-NMR, ^{1}H-NMR and IR spectroscopy

Chemical & Safety Data

Substance	MW /(g/mol)	m.p./ °C	b.p./ °C	Density/(g/mL)	Function
Dichloromethane CH_2Cl_2	84.93	-97	39~40	1.32	Solvent
	Safety: Irritant, possible cancer hazard, inhalation may cause CNS effects				
Sodium sulfate Na_2SO_4	142.04	884	N/A	N/A	Drying agent
	Safety: Possible irritant				
Eugenol $C_{10}H_{12}O_2$	164.20	-12 ~ -10	254	1.06	Analyte
	Safety: Irritant, potential allergen				
Chloroform-d $CDCl_3$	120.38	-64	61	1.5	Solvent
	Safety: Irritant, possible cancer hazard, inhalation may cause CNS effects				

Safety Notes

1. When heating a reaction apparatus, be sure that it is open to the air so that pressure build up and rupture of the apparatus does not occur.

2. When heating liquids, make sure the liquid is stirred (or a boiling chip is added) to prevent "bumping".

3. When performing an extraction, make sure to vent the separatory funnel often to prevent pressure build-up.

Note: N/A means not applicable and CNS effects stand for central nervous system effects.

Experiment datasheet

Course: CHM192 Organic Experiment (I)   Date: 04 / 28 / 2021
Name: Wang Xiaolun
Title: Isolation of Eugenol by Steam Distillation of Cloves

1. Apparatus for steam distillation of eugenol show in Figure 1. The apparatus sketched below was assembled and charged with 5.032 g of ground cloves and 40 mL distilled water. The cloves were soaked for about 15 min until thoroughly wetted.

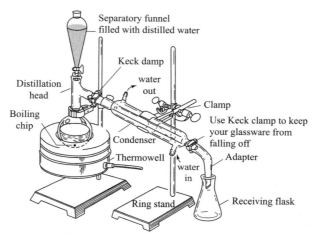

Figure 1   Apparatus for steam distillation of eugenol

2. The mixture was heated to boiling. Initial hot plate setting = 3. After 20 min, mixture was still not boiling so setting increased to 7. Distillate collected at rate of about 1 drop/2~3 sec. & about ~~20.0 mL~~ 30.5 mL of distillate collected, then distillation was discontinued.

3. The distillate was transferred to a separatory funnel, extracted with 2 mL dichloromethane, then again with (2× 1 mL) dichloromethane.

4. The combined organic extracts were dried over sodium sulfate, and transferred to a tared ~~beaker~~ conical vial (Tare weight= 18.643 g).

5. The dichloromethane was evaporated by heating on a hot plate (IN FUME HOOD!) to yield a pale yellow oil. Vial+ eugenol= 19.028 g, and the mass of eugenol obtained= 0.385 g.

   Recovery/% = (0.385 g/5.032 g) ×100= 7.65 %

6. A mass spectrum, a ^{1}H-NMR spectrum, a ^{13}C-NMR spectrum and an IR spectrum were obtained.

Lab report

Course CHM 192 Organic Experiment (Ⅰ)
Name Wang Xiaolun
Student ID 20210123
Title Isolation of Eugenol from Cloves by Steam Distillation and its Identification by mass, ^{1}H-NMR, ^{13}C-NMR, and Infrared Spectroscopy
Date 28th April, 2021
Purpose To isolate eugenol from cloves by steam distillation and analyze it by using mass, ^{13}C-NMR, ^{1}H-NMR and IR spectroscopy
Principle[1,2]

"Essential oils" are the volatile components associated with the aromas of many plants. In this experiment, the essential oil eugenol (the main component of oil of cloves) will be isolated from ground cloves by steam distillation and identified by mass, NMR and infrared spectroscopy.

The principle of steam distillation is based on the fact that two immiscible liquids will boil at a lower temperature than the boiling points of either pure component, because the total vapor pressure of the heterogeneous mixture is simply the sum of the vapor pressures of the individual components (i.e. $P_T = P_1^\ominus + P_2^\ominus$, where $P^\ominus$ is the vapor pressure of the pure liquids). This leads to a higher vapor pressure for the mixture than would be predicted for a solution using Raoult's Law (that is $P_T = P_1^\ominus N_1 + P_2^\ominus N_2$, where N is the mole fraction of the component in the mixture). The higher total vapor pressure leads to a lower boiling point for the mixture than for either single component.

When two immiscible liquids are distilled, the total vapor pressure P_T above the liquid is equal to the sum of the vapor pressures of each compound. This relationship, known as Dalton's law: $P_T = P_1^\ominus + P_2^\ominus$. The respective mole fractions are not included in this equation, because, in an ideal situation, each liquid vaporizes independently of the other. When P_T is equal to atmospheric pressure of 760 torr (1 torr = 1 mmHg), compounds 1 and 2 begin to codistill, with each compound contributing to P_T.

Consider water as compound 1. The vapor pressure of pure water at its boiling point of 100 ℃ is 760 torr. Because compound 2 also contributes to P_T, the mixture will distill at a temperature less than 100 ℃. The actual distillation temperature will depend on the vapor pressure of compound 2. Steam distillation offers an advantage in that volatile compounds that are unstable or have high boiling points can codistill with water at relatively low temperatures. This process avoids decomposition that might occur at the normal boiling point of the compound of interest. For example, eugenol, the major compound of clove oil, boils at a relatively high temperature of 254 ℃. Steam distillation avoids this high temperature and results in the distillation of eugenol at a temperature slightly less than 100 ℃.

Steam distillation can be carried out in two ways: the direct method and the live steam method. In the direct method, steam is generated by boiling a mixture of the source of the compound of interest and water. The live steam method is carried out by passing steam from an external source into the distillation flask. The direct method of steam distillation will be used in this experiment.

In practice, steam distillation is usually carried out by one of two methods. In the first method, an excess of water is added to the compound of interest in a distilling flask. The mixture is then heated to the boiling point. The resulting vapor is condensed and collected in a receiving flask. The compound of interest is then separated from the water, often by extraction. In the second method, steam is bubbled into the compound of interest to effect the distillation. In this experiment, you will use the first method because it is easier to set up.

Clove oil belongs to a large class of natural products called the **essential oils**. Many of these compounds are used as flavorings and perfumes and, in the past, were considered to be the "essence" of the plant from which they were derived. Cloves are the dried flower buds of the clove tree, Eugenia caryo-phyllata, found in India and other locations in the Far East. Steam distillation of freshly ground cloves results in clove oil, which is a mixture of several compounds. Eugenol is the major compound, comprising 85% ~ 90%. Eugenol acetate comprises 9% ~ 10 %. These structures are shown below. Some physical constants of reagents used in this experiment are listed in Table 1.

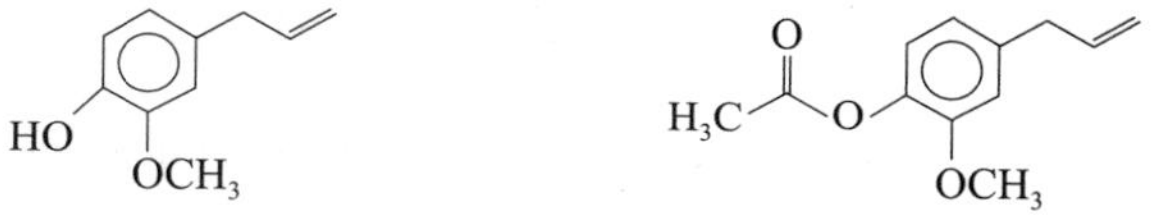

eugenol

eugenol acetate

Table 1 Reagents and physical constants

Substance	Quantity used	MW/(g/mol)	m.p. / °C	b.p. / °C	Density/(g/mL)
Dichloromethane	21 mL	84.93	−97	39~40	1.32
Sodium sulfate	0.5 g	142.04	884	N/A	N/A
Eugenol	product	164.20	−12~−10	254	1.06
Chloroform-d		120.38	−64	61	1.5
Cloves, ground	5.032 g				
Sodium chloride, sat. solution	10 mL	58.44			

Apparatus

A thermowell (heating mantle); 10 mL, 100 mL graduated cylinders; steam distillation glasswares;

A 50 mL pipet; 50 mL, 100 mL Erlenmeyer flask; an addition 30 mL separatory funnel; NMR tube;

IR plates; tin foil.

Experimental

Part One: Steam Distillation of Eugenol[3,4]

The apparatus was assembled as shown in Figure 1. The 250 mL round bottom flask was charged with 5.032 g of ground cloves and 40 mL of distilled water. A boiling chip was added into the flask. A 100-mL Erlenmeyer flask was used as a receiver. 25 mL of distilled water was poured into the addition 30-mL separatory funnel. The cloves were allowed to soak in the water until thoroughly wetted (about 15 minutes). The mixture was heated gently.

Foam would be formed if the clove oil was heated rapidly. The mixture was kept boiling and steam distilled. The distillate was collected at the rate of about one drop every 3~4 s. Distilled water was added to the 250-mL round bottom flask to keep the water level. After about 30 mL of distillate had been collected, the steam distillation was stopped.

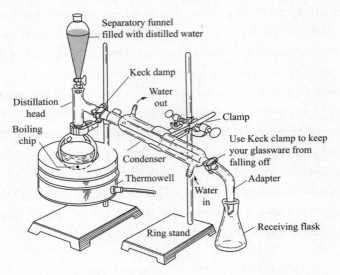

Figure 1 Apparatus for steam distillation of eugenol

Part Two: Extraction of Clove Oil

Caution: Dichloromethane is a strong irritant. This portion of the experiment must be done in a fume hood. Clove oil (eugenol) is irritating. Prevent eye, skin, and clothing contact.

Using a pipet, the distillate was transferred from the receiver into a 125-mL separatory funnel. 10 mL of saturated NaCl solution was added. The condenser was dismantled from the distillation head. Significant amounts of clove oil were adhered on the condenser and the sides and neck of the receiving flask. Using a dropper pipet, the condenser and the inside neck of the receiving flask were carefully rinsed with 5 mL of dichloromethane. The flask was gently swirled to dissolve the remaining clove oil. The dichloromethane washings were transferred to the distillate in the separatory funnel using a pipet. The separatory funnel was capped and the contents were gently swirled for several seconds. The separatory funnel was vented frequently. After the pressure has been vented, the contents were shaken vigorously to thoroughly mix the two layers. The separatory funnel was swirled. At the same time, the outside of the separatory funnel was gently tapped with the index finger to force any droplets of dichloromethane that are adhering to the sides of the funnel into the bottom layer. The layers were allowed to separate. The stopper was removed and the lower dichloromethane layer was drained into a 50-mL Erlenmeyer flask. A second 5-mL portion of dichloromethane was added into the separatory funnel. Transfer the rinsing to the separatory funnel. The extraction procedures of the aqueous layer were repeated. The second dichloromethane extract was drained from the separatory funnel and combined with the first one in the 50-mL Erlenmeyer flask. The rinsing and extraction process were repeated with a third 5-mL portion of dichloromethane. The third extract was combined in the same 50-mL Erlenmeyer flask.

Part Three: Drying the Dichloromethane Extracts

Caution: Anhydrous sodium sulfate (Na_2SO_4) is irritating and hygroscopic. Do not inhale and ingest this compound. This portion of the lab must be done in the fume hood.

Approximately 0.5 g increments of anhydrous Na_2SO_4 were added to the Erlenmeyer flask containing the dichloromethane extracts until the anhydrous Na_2SO_4 had no longer formed clumps. The flask was covered with a piece of tin foil. The extracts were allowed to dry for 5 minutes.

A clean, dry 50-mL round bottom flask was weighed to the nearest 0.001 g and the mass was recorded. Using a pipet, the dried dichloromethane was transferred into the preweighed round bottom flask carefully, making certain that no Na_2SO_4 was transferred with the solution. Three additional 2-mL portions of dichloromethane were used to rinse the Na_2SO_4 and complete transfer of the clove oil to the round bottom flask could be achieved. A piece of boiling chip was added into the flask. The dichloromethane extracts were evaporated on a rotary evaporator. When all of the dichloromethane had been evaporated, the round bottom flask was covered with a piece of tin foil and cooled to room temperature. The clove oil and the flask were weighed to the nearest 0.001 g and the total mass was recorded. The mass of the clove oil was obtained by subtracting the mass of the empty flask.

Result (Data and Analysis)

Mass of eugenol isolated = 0.385 g

0.385 g of eugenol was recovered from 5.032 g of cloves.

Percentage recovery = Mass of eugenol isolated/ Mass of cloves used
= 0.385 g / 5.032 g × 100 %
= 7.65 %

Spectroscopic Identification of Eugenol[5]

The mass spectrum, ^{1}H-NMR spectrum, ^{13}C-NMR spectrum and IR spectrum are shown in Figure 2~Figure 5 respectively.

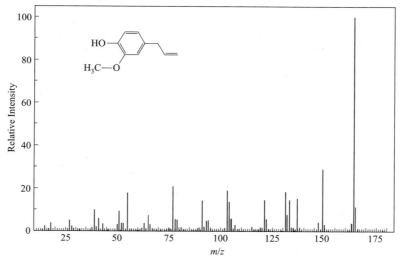

Figure 2 Mass spectrum of eugenol

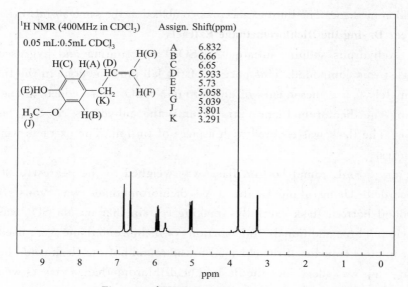

Figure 3 ^{1}H-NMR spectrum of eugenol

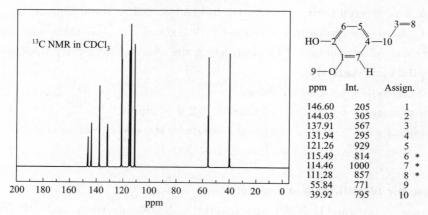

Figure 4 ^{13}C-NMR spectrum of eugenol

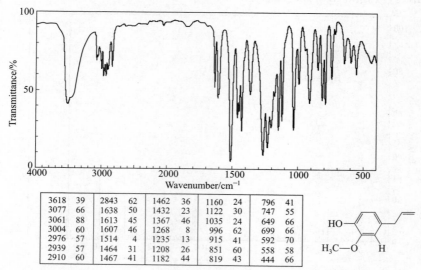

3618	39	2843	62	1462	36	1160	24	796	41
3077	66	1638	50	1432	23	1122	30	747	55
3061	88	1613	45	1367	46	1035	24	649	66
3004	60	1607	46	1268	8	996	62	699	66
2976	57	1514	4	1235	13	915	41	592	70
2939	57	1464	31	1208	26	851	60	558	58
2910	60	1467	41	1182	44	819	43	444	66

Figure 5 IR spectrum of eugenol

IR Data of Eugenol

FT-IR (film, NaCl plates): 3560 (OH), 3080-3000 (sp^2 CH), 2980-2940 (sp^3 CH), 1640 (alkene C=C), 1600, 1514 (aromatic C=C) cm^{-1}.

Discussion

Steam distillation of cloves produced 0.0770 g of an oil which contained in its IR spectrum the functional groups O—H (at 3560 cm^{-1}), sp^2 C—H (3080-3000 cm^{-1}), aliphatic C—H (2980-2940 cm^{-1}), and both alkene C=C (at 1640 cm^{-1}) and aromatic C=C (at 1600 and 1514 cm^{-1}). The IR spectrum is attached to this report. These data are consistent with the structure of eugenol, shown below

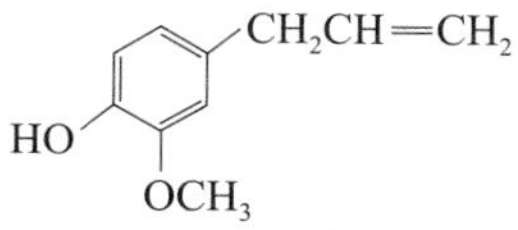

$$\text{HO} - \text{C}_6\text{H}_3(\text{OCH}_3)(\text{CH}_2\text{CH}=\text{CH}_2)$$

In addition, the IR spectrum of the product from the steam distillation of cloves closely corresponds with that of an authentic sample of eugenol. Therefore, it can be concluded that the oil which was isolated from cloves is in fact, eugenol.

Although the recovery (7.46 %) seems slightly low relative to the expected 10%[2], the experiment proceeded as planned. There were no spills or other abnormal physical losses. It is possible that the ratio of the size of the glassware to the theoretical amount of eugenol which can be obtained from cloves in this experiment is large, leading to adherence of a large percentage of the product on the sides of the glass apparatus. If this is so, then steam distillation of a larger sample of cloves should give an improved recovery. Otherwise, it can be concluded that the specific sample of cloves used contains approximately 7.5 % eugenol.

Conclusion

In this experiment, it was shown that about 7.5% of eugenol could be recovered from cloves by steam distillation. This oil was identified as eugenol by comparison of its mass, ^{1}H-NMR, ^{13}C-NMR and infrared spectra with the authentic sample.

References

[1] Pavia D L, Lampman G M, Kriz G S, et al. Introduction to Organic Laboratory Techniques, A Microscale Approach; 3rd ed.; Brooks/Cole: Pacific Grove, CA, 1999: 139, 663, 665.

[2] Guan W, Li S, Yan R, et al. Comparison of essential oils of clove buds extracted with supercriticalcarbon dioxide and other three traditional extraction methods [J]. Food Chemistry, 2007, 101: 1558-1564.

[3] Fieser L F, Williamson K L. Organic Experiments, 7th ed. D. C. Heath and Company, USA, 1992: 66.

[4] Fieser L F, Williamson K L. Organic Experiments, 7th ed. D. C. Heath and Company, USA, 1992: 78-80.

[5] The MS, ^{1}H-NMR, ^{13}C-NMR and FT-IR spectra of eugenol are obtained from https://spectrabase.com/spectrum/GICLnD5zWZM.

7.2 Some Guidelines for Writing Scientific Papers

7.2.1 Abstract

The first item to appear after the title of the document is the abstract (sometimes called the summary or executive summary). It is a very concise summary of all the salient aspects of the entire document. An abstract is written so that a reader interested in the work, can gather an impression of the contents of the report and decide if investigating the details further is worthwhile.

It should include:
(ⅰ) The aim of the experiment.
(ⅱ) The background context.
(ⅲ) The procedures followed and equipment used.
(ⅳ) The results that were obtained.
(ⅴ) Any observations made.
(ⅵ) The findings drawn and the impact those findings have towards fulfilling the original aim.

Compressing all this information is a very short piece of text makes writing an abstract a difficult task to perform and one that is often done badly by undergraduate students. Practice writing abstracts is one of the best methods for improving technique.

There are some rules that should be followed when writing an abstract:
(ⅰ) The structure of an abstract should follow the structure of the report.
(ⅱ) Only the critically important "headlines" from the report should be included.
(ⅲ) It shouldn't include tables, graphs, pictures or equations.
(ⅳ) It should be self-contained, i. e. can be read and understood without needing to refer to other documents.
(ⅴ) It should not include abbreviations, acronyms or jargon.
(ⅵ) It is the first thing to appear after the title of the document, but should be the last part of the document to be written.

7.2.2 Introduction

In a scientific research article, this section of the paper is devoted to making the case for why the work is important and significant and for discussing the previous work reported in the literature that has led up to the work being reported in this paper.

It answers the questions: What problem has been addressed? What is the significance of the problem? What theory or principle behind it?

The introduction provides the reader with the background to the work documented in the report. This section should set the scene for what is to follow. It should contain the aims or objectives of the proposed work. If an aim of the experiment is to investigate a hypothesis, this should be stated in the introduction. The aims, objectives and/or hypothesis should be given in the context of the real world application outside the experiment.

There should be a broad introduction to the background of the science, the reasons for doing the work, and who will benefit from the results. It is not necessary to derive equations from first principles or exhaustively describe theory, as reference to alternative material can point the reader towards where to find this information. Given that the reader may not be familiar with the specifics of the discipline, it may be necessary to explain acronyms or technical terms. This should be done in the introduction. In addition, there should be a summary of previously conducted work in the same field and a description of how the contents of the report furthers the advancement of knowledge.

In summary, the introduction should include:
(i) A background to the subject.
(ii) Previously conduced work in the same subject.
(iii) Aims and objectives for the work that will be presented in the lab report.
(iv) Reasons why the work is being conducted.

7.2.3 Experimental Methods

This section should provide the details of how the experiment was carried out. It should not be written as a recipe but more as a journal entry; a fairly detailed account of what was done in lab. A description of how the data was processed should also be part of the methods section.

In addition, the procedure section is a record of what was done, a chronological description of the steps followed and the equipment used. It should not be a list of instructions but should be written as prose, in the third person and past tense (as should the rest of the report). Details of what variables were recorded, what observations were made, and what types of instrumentation were used should be included.

The procedure should contain sufficient detail to allow the experiment to be repeated by another person at a later date. It is necessary to give a detailed record of any important conditions of the experiment (e.g. operating temperatures, atmospheric pressure, humidity), any specific techniques that were used (e.g. equipment calibration) and any materials involved (e.g. 10mol/L hydrochloric acid, cast iron). It is critical to include all relevant information but to ensure the report is sufficiently concise and excludes extraneous detail. For example, in detailing equipment, it may be useful to record the manufacturer and model number, the precision of the instrument, the zero-offset, any calibration that was performed, and the accuracy at different recording ranges. A record of the color of the equipment will probably be of no consequence to the results and should not be included in the report.

A well written procedure should include not only a description of what was performed, but also the reasoning behind the experimental design. Why was the experiment set up in the way it was and how does it conform to the scientific method? What special measures have been put in place to ensure accuracy and repeatability of the results?

To describe equipment, labelled diagrams and photographs can be included. Photographs are usually not sufficient to replace explanatory diagrams, and should only be used if they enhance readers' understanding of the experiment. Any safety precautions or procedures that were observed, or any PPE (personal protective equipment) that was used can be discussed if appropriate.

7.2.4 Results (Data and Analysis)

Suggestions:
(ⅰ) Tabulate the raw data, plot the spectra obtained directly from the experiments.
(ⅱ) All data, raw or final should be with proper units and proper error.
(ⅲ) All tables and graphs should be clearly titled and labeled.

The results section of a lab report contains an impartial description of the results obtained from the experiment, typically presented as tables or graphs, and observations that were made. At this point in the report, interpretation of the results should not be performed. To convey the main findings of the experiment, processed, rather than raw data, should be shown.

A brief description of the method used to covert the raw data to the results could be included, possibly using an illustrative sample calculation. However, large datasets and numerous intermediate calculations should not be shown in the results section. These can be included for reference in an appendix if useful for the reader. Large quantities of raw data can be stored electronically and an explanation of how to access it given in the report.

In addition to the measurements taken during the experiment, the results section should include any observations that were made during the experiment. Unexpected phenomena may affect the results in ways that are not known by the author of the lab report but may be of significance to the reader. For example, if work is conducted on a water flow system and a large number of bubbles are observed in the supply or there is a large oscillation in the values reported from measurement equipment, record this in the results section.

A well written results section of a lab report highlights the trends observed rather than giving details of exact results. The data presented in the results section should demonstrate how the experiment's objectives have been met. For example, if the aim of an experiment was to optimize the level of fuel consumption in a petrol car by varying travelling speed, then the results section could show a plot of kilometers per liter against meters per second. The details of the amount of fuel used, distance travelled by the car, the variation of lengths of journeys, the elimination of effects of acceleration and deceleration on the results, and other processing techniques should only be described briefly.

7.2.5 Discussion

In some papers, these are combined and in other they form two different sections. We will keep them separate in this course. The discussion section will discuss the significance of the findings from the data analysis section. It is also in this section the questions that are being asked are addressed in the context of a well-written paragraph.

The purposed of a discussion section is to answer the questions:

(ⅰ) What do the results mean?

(ⅱ) Do they answer the questions the experiment was to investigate?

(ⅲ) What is the relevance to engineering problems?

(ⅳ) Where are errors introduced?

The discussion section is used to analyse and interpret the information presented in the results section. Mention should be made of whether or not the results achieve the aims or prove/disprove the hypothesis previously set out, within the context of the background science. In doing so, the discussion should refer to the introduction section so that the document is a coherent piece of work.

The interpretation of the results should discuss the physical principles for the trends or phenomena that were observed. If unexpected results are produced, that were not suggested from the background theory presented in the introduction, the discussion allows possible reasons for these findings to be proposed.

The discussion section should attempt to report on the errors and uncertainties in the experiment. Errors may include the limited precision of instruments, the result of ignoring wind resistance, or human error/reaction time. Where possible these errors should be quantified, even approximately, and ranked. Further details on handling and manipulating errors are given in this document.

There are two reasons to quantify errors and uncertainties:

(ⅰ) It allows a degree of confidence to be placed on the results presented.

(ⅱ) It allows efforts to reduce error in future experiments to be focused correctly.

The potential impact of the results on the real world applications to which the experiment was designed to apply should be discussed in this section. For example, by: comparison between field scale and lab scale results; proposing design changes to existing products based on new knowledge; quantifying the impact of the results on beneficiaries.

7.2.6 Conclusion

This section should consist of at least two paragraphs that summarize what happened in the experiment, and what you learned while doing the experiment. State whether your hypothesis (prediction) was accepted or rejected, and what your data and results mean. Please remember that your interpretation of the experiment and its results is the most important part of your lab report.

The conclusion is a short review of that which has been deduced from the work conducted. It is an opportunity to restate the aims or key questions and to summarize the key points raised in the results and discussion sections. No new information should be given in the conclusion that hasn't been stated previously in the document.

Proposals for further work or potential improvements identified during the experiment can be suggested in the conclusion, or this can be placed in a separate "further work" section following the conclusion.

7.2.7 References

This section lists the source of non-routine information. Actually, following the conclusion should be additional, non-essential information. Any previously published work cited in the body of the document should be referenced in a dedicated "references" section. If previously published work has been used but not explicitly cited, this should be placed in a "bibliography" section.

Other information, such as raw data, manufacture's user manuals, complex numerical tables of results…etc. can be placed in an appendix, to which the reader can refer for detail. The appendix is not a substitute for the results section and important information must be in the body of the report.

A more specific guideline for preparation of figures and tables, and instructions on how to submit your paper to different journals and publishers, students are suggested to read the Guide, Notes, or Instructions for Authors that appear in each online publication's first issue of the year.

You may find some useful information and resources from the following websites:

(1) The American Chemical Society: https://pubs.acs.org
(2) The Royal Society of Chemistry: https://www.rsc.org
(3) Elsevier Science Direct: http://www.sciencedirect.com
(4) Wiley Online Library: https://onlinelibrary.wiley.com
(5) Taylor & Francis: http://www.tandfonline.com
(6) Thieme-Chemistry: http://www.thieme-chemistry.com/thieme-chemistry/journals
(7) Sci-Finder: https://sso.cas.org
(8) Web of Science SCI (Science Citation Index): http://www.webofknowledge.com
(9) Springer Protocol: https://experiments.springernature.com
(10) Nature: https://www.nature.com
(11) Science: https://www.sciencemag.org

Appendices

Appendix A Names of Chemical Elements

Atomic Number	Symbol	Element (English)	Element (Chinese)
1	H	Hydrogen	氢
2	He	Helium	氦
3	Li	Lithium	锂
4	Be	Beryllium	铍
5	B	Boron	硼
6	C	Carbon	碳
7	N	Nitrogen	氮
8	O	Oxygen	氧
9	F	Fluorine	氟
10	Ne	Neon	氖
11	Na	Sodium	钠
12	Mg	Magnesium	镁
13	Al	Aluminum	铝
14	Si	Silicon	硅
15	P	Phosphorus	磷
16	S	Sulfur	硫
17	Cl	Chlorine	氯
18	Ar	Argon	氩
19	K	Potassium	钾
20	Ca	Calcium	钙
21	Sc	Scandium	钪
22	Ti	Titanium	钛
23	V	Vanadium	钒
24	Cr	Chromium	铬
25	Mn	Manganese	锰

续表

Atomic Number	Symbol	Element (English)	Element (Chinese)
26	Fe	Iron	铁
27	Co	Cobalt	钴
28	Ni	Nickel	镍
29	Cu	Copper	铜
30	Zn	Zinc	锌
31	Ga	Gallium	镓
32	Ge	Germanium	锗
33	As	Arsenic	砷
34	Se	Selenium	硒
35	Br	Bromine	溴
36	Kr	Krypton	氪
37	Rb	Rubidium	铷
38	Sr	Strontium	锶
39	Y	Yttrium	钇
40	Zr	Zirconium	锆
41	Nb	Niobium	铌
42	Mo	Molybdenum	钼
43	Tc	Techetium	锝
44	Ru	Ruthenium	钌
45	Rh	Rhodium	铑
46	Pd	Palladium	钯
47	Ag	Silver	银
48	Cd	Cadmium	镉
49	In	Indium	铟
50	Sn	Tin	锡
51	Sb	Antimony	锑
52	Te	Tellurium	碲
53	I	Iodine	碘
54	Xe	Xenon	氙
55	Cs	Cesium	铯
56	Ba	Barium	钡
57	La	Lanthanum	镧
58	Ce	Cerium	铈
59	Pr	Praseodymium	镨
60	Nd	Neodymium	钕
61	Pm	Promethium	钷
62	Sm	Samarium	钐
63	Eu	Europium	铕
64	Gd	Gadolinium	钆
65	Tb	Terbium	铽
66	Dy	Dysprosium	镝
67	Ho	Holmium	钬
68	Er	Erbium	铒
69	Tm	Thulium	铥
70	Yb	Ytterbium	镱
71	Lu	Lutetium	镥
72	Hf	Hafnium	铪

续表

Atomic Number	Symbol	Element (English)	Element (Chinese)
73	Ta	Tantalum	钽
74	W	Tungsten	钨
75	Re	Rhenium	铼
76	Os	Osmium	锇
77	Ir	Iridium	铱
78	Pt	Platinum	铂
79	Au	Gold	金
80	Hg	Mercury	汞
81	Tl	Thallium	铊
82	Pb	Lead	铅
83	Bi	Bismuth	铋
84	Po	Polonium	钋
85	At	Astatine	砹
86	Rn	Radon	氡
87	Fr	Francium	钫
88	Ra	Radium	镭
89	Ac	Actinium	锕
90	Th	Thorium	钍
91	Pa	Protactinium	镤
92	U	Uranium	铀
93	Np	Neptunium	镎
94	Pu	Plutonium	钚
95	Am	Americium	镅
96	Cm	Curium	锔
97	Bk	Berkelium	锫
98	Cf	Californium	锎
99	Es	Einsteinium	锿
100	Fm	Fermium	镄
101	Md	Mendelevium	钔
102	No	Nobelium	锘
103	Lw	Lawrencium	铹
104	Rf	Rutherfordium	𬬻
105	Db	Dubnium	𬭊
106	Sg	Seaborgium	𬭳
107	Bh	Bohrium	𬭛
108	Hs	Hassium	𬭶
109	Mt	Meitnerium	鿏
110	Ds	Darmstadtium	𫟼
111	Rg	Roentgenium	𬬭
112	Cn	Copernicium	鿔
113	Nh	Nihonium	𬭊
114	Fl	Flerovium	𫓧
115	Mc	Moscovium	镆
116	Lv	Livermorium	𫟷
117	Ts	Tennessine	䃲
118	Og	Oganesson	鿫

Appendices

Appendix B   Common Chemical Terms

(1) Matter

Accuracy（准确度）is the closeness of a measured value to the true or accepted value of a quantity.

The **atom**（原子）is an extremely small particles of matter that retains its identity during chemical reactions.

Chemical change（化学变化）is a process in which one or more substances are changed into one or more substances.

A **chemical process**（化学过程）is a process in which the identity of matter changes.

The **chemical property**（化学性质）is a characteristic of a material involving its chemical change.

A **chemical reaction**（化学反应）is a process in which one set of substances (reactants) is transformed into a new set of substances (products).

A **compound**（化合物）is a substance made up of two or more elements chemically combined.

Composition（组成）refers to the components and their relative proportions in a sample of matter.

An **element**（元素）is a substance composed of a single type of atom. It cannot be broken down into simpler substances by chemical reactions.

The **extensive property**（广度性质）is one, like mass or volume, whose values depend on the quantity of matter observed.

A **homogeneous mixture**（均相混合物）is a mixture of elements and/or compounds that has a uniform composition and properties within a given sample.

A **heterogeneous mixture**（多相混合物）is a mixture that consists of physically distinct parts, each with different properties.

The **intensive property**（强度性质）is independent of the quantity of matter involved in the observation.

The **law of constant composition**（定比定律）states that all samples of a compound have the same composition, that is, the same proportions by mass of the constituent elements.

Matter（物质）is anything that occupies space and has mass.

A **mixture**（混合物）is a combination of two or more substances in which the substances retain their distinct identities.

A **molecule**（分子）is a combination of two or more atoms in a specific arrangement held together by chemical bonds.

A **phase**（相）is one of several different homogeneous materials presented in the portion of matter under study.

Precision（精密度）is the degree of reproducibility of a measured quantity- the closeness of agreement among repeated measurements.

A **qualitative property**（定性性质）is a property of a system that can be determined by general observation.

A **quantitative property**（定量性质）is a property of a system that can be measured and expressed with a number.

A **substance**（物质）has a constant composition and properties throughout a given sample and from one sample.

A **theory**（理论）is a tested explanation of nature basic phenomena.

(2) Atoms and the Periodic Table

Alkali metal（碱金属）is the family name for the group 1A elements of the periodic table.

Alkaline earth metal（碱土金属）is the family name for the group 2A elements of the periodic table.

The **atomic mass (weight)**（原子量）of an element is the average of the isotopic masses weighted according to the naturally occurring abundances of the isotopes of the element and relative to the value of exactly 12 amu for a carbon-12 atom.

An **atomic mass unit**（原子质量单位），amu，is a mass unit equal to exactly one-twelfth the mass of a carbon-12 atom.

The **atomic number**（原子序数），Z，is the number of protons in the nucleus of an atom. It is also the number of electrons outside the nucleus of an electrically neutral atom.

The **d-block**（d 区）refers to that section of the periodic table in which the process of orbital filling (aufbau process) involves a d subshell.

An **electron**（电子）is a negatively charged subatomic particle found outside the nucleus of all atoms.

A **family**（族）of elements is a numbered group from the periodic table, sometimes carrying a distinctive name.

The **f-block**（f 区）is that portion of the periodic table where the process of filling of electron orbitals (aufbau process) involves subshells. These are the lanthanide and actinide elements.

A **group**（族）is a vertical column of elements in the periodic table. Members of a group have similar properties.

Halogens（卤素）（group 7A）are the most reactive nonmetals, having the electron configuration in the electronic shell of highest principal quantum number.

Isotopes（同位素）of an element are atoms with different numbers of neutrons in their nuclei. That is, isotopes of an element have the same atomic numbers but different mass numbers.

The **law of multiple proportions**（倍比定律）states that if two elements form two or more compounds, the masses of one element combined with a fixed mass of the second are in the ratio of small whole numbers when the different compounds are compared.

The **lanthanides**（镧系元素）are the elements ($Z = 58 \sim 71$) characterized by a partially filled subshell in their atoms. Because lanthanum resembles them, La ($Z = 51$) is generally considered together with them.

The **main-group elements**（主族元素）are those in which s or p subshells are being filled in the aufbau process. They are also referred to as the s-block and p-block elements. They are found in groups 1, 2, and 13~18 in the periodic table (the A groups).

The **mass number**（质量数），A, is the total of the number of protons and neutrons in the nucleus of an atom.

A **mass spectrometer**（质谱仪）is a device used to separate and to measure the quantities and masses of different ions in a beam of positively charged gaseous ions. It does not change its identity in physical changes, but it can be broken down into its constituent elements by chemical change.

A **metal**（金属）is a substance or mixture that has a characteristic luster or shine, is generally a good conductor of heat and electricity, and is malleable and ductile.

A **metalloid**（类金属）is an element that may display both metallic and nonmetallic properties under the appropriate conditions.

A **neutron**（中子）is an electrically neutral subatomic particle with a mass slightly greater than that of a proton.

Noble gases（稀有气体）are elements whose atoms have the electron configuration in the electronic shell of highest principal quantum number.

A **nonmetal**（非金属）is an element that does not exhibit the characteristics of a metal.

A **nucleus**（原子核）is the central core of the atom that contains the protons and neutrons.

A **nuclide**（核素）is a particular atom characterized by a definite atomic number and mass number.

The **p-block**（p 区）is that portion of the periodic table in which the filling of electron orbitals (aufbau process) involves p subshells.

Percent natural abundances（自然丰度百分比）refer to the relative proportions, expressed as percentages by number, in which the isotopes of an element are found in natural sources.

A **proton**（质子）is a positively charged particle in the nucleus of an atom.

The **s-block**（s 区）refers to the portion of the periodic table in which the filling of electron orbitals (aufbau process) involves the s subshell of the electronic shell of highest principal quantum number.

Transition elements（过渡元素）are those elements whose atoms feature the filling of a d or subshell of an inner electronic shell. If the filling of a subshell occurs, the elements are sometimes referred to as inner transition elements.

(3) Chemical Compounds

An **anion**（阴离子）is a negatively charged ion.

Binary compounds（二元化合物）are compounds composed of two elements.

A **cation**（阳离子）is a positively charged ion.

A **chemical formula**（化学式）is a notation that uses atomic symbols with numerical subscripts to convey the relative proportions of atoms of the different elements in a substance.

An **empirical formula**（实验式）is the simplest chemical formula that can be written for a compound, that is, having the smallest integral subscripts possible.

A **formula unit**（化学式单位）is the smallest collection of atoms or ions from which the empirical formula of a compound can be established.

The **formula weight**（式量）is the sum of the atomic weights of all atoms in a formula unit of a compound.

A **hydrate**（水合物）is a compound in which a fixed number of water molecules is associated with each formula unit, such as $CuSO_4 \cdot 5H_2O$.

An **ion**（离子）is a charged species consisting of a single atom or a group of atoms. It is formed when a neutral atom or a covalently bonded group of atoms either gains or loses electrons.

An **ionic compound**（离子化合物）is a compound composes of cations and anions.

A **molecular compound**（分子化合物）is a compound comprised of discrete molecules.

A **molecular formula**（分子式）denotes the numbers of the different atoms present in a molecule. In some cases, the molecular formula is the same as the empirical formula; in others it is an integral multiple of that formula.

A **monoatomic ion**（单原子离子）is an ion formed from a single atom.

An **oxoacid**（含氧酸）is an acid containing hydrogen, oxygen, and another elements (often called the central atom).

An **oxoanion**（含氧阴离子）is a polyatomic anion containing a nonmetal, such as Cl, N, P, or S, in combination with some number of oxygen atoms.

A **polyatomic ion**（多原子离子）is a combination of two or more chemically bonded atoms that exists as an ion.

Ternary compounds（三元化合物）are comprised of three elements.

(4) Chemical Reactions

The **actual yield**（实际产量）is the measured quantity of a product obtained in a chemical reaction.

A **balanced equation**（配平方程式）has the same number of atoms of each type on both sides.

By-products（副产物）are substances produced along with the principal product in a chemical process, either through the main reaction or a side reaction.

A **chemical equation**（化学方程式）is a symbolic representation of a chemical reaction in terms of chemical formulas.

A **chemical reaction**（化学反应）is a process that neither creates nor destroys atoms but rearranges atoms in elements and/or chemical compounds.

A **combination reaction**（化合反应）is a reaction in which two or more reactants combine to form a single product.

A **combustion reaction**（燃烧反应）is a reaction in which a substance burns in the presence of oxygen.

Consecutive reactions（连串反应）are two or more reactions carried out in sequence. A product of each reaction becomes a reactant in a following reaction until a final product is formed.

A **decomposition reaction**（分解反应）is a reaction in which two or more products form from a single reactant.

An **ionic equation**（离子方程式）is a chemical equation in which all strong electrolytes are shown as ions. Also known as **complete ionic equation**（完全离子方程式）.

The **law of conservation of mass**（质量守恒定律）states that the total mass of the products of a chemical reaction is the same as the total mass of the reactants entering into the reaction.

The **limiting reactant**（**reagent**）（限量反应物）in a reaction is the reactant that is consumed completely. The quantity of product（s）formed depends on the quantity of the limiting reactant.

A **metathesis reaction**（**double replacement reaction**）（复分解反应）is a reaction between two salts, symbolized EF and GH, that switch cationic partners to form two different compounds, symbolized EH and GF.

A **molecular equation**（分子方程式）is a chemical equation written with all compounds represented by their chemical formulas.

A **net ionic equation**（净离子方程式）is a chemical equation from which spectator ions have been removed.

The **percent yield**（产率）is the percent of the theoretical yield of product that is actually obtained in a chemical reaction.

A **precipitate**（沉淀物）is an insoluble solid product that separates from a solution.

A **precipitation reaction**（沉淀反应）is a chemical reaction in which a precipitate forms.

A **product**（产物）is a substance that is formed in a chemical reaction.

A **reactant**（反应物）is a substance that is consumed in a chemical reaction.

A **side reaction**（副反应）is a reaction that produces an undesired or unexpected product and accompanies a reaction intended to produce something else.

Simultaneous reactions（平行反应）are two or more reactions that occur at the same time.

Spectator ions（旁观离子）are ionic species that are present in a reaction mixture but do not take part in the reaction.

The **theoretical yield**（理论产量）is the quantity of product calculated to result from a chemical reaction.

Stoichiometric coefficients（计量系数）are the coefficients used to balance an equation.

Stoichiometry（化学计量学）refers to quantitative measurements and relationships involving substances and mixtures of chemical interest.

(5) Gases

Avogadro's law（**hypothesis**）（阿伏伽德罗定律）states that at a fixed temperature and pressure, the volume of a gas is directly proportional to the amount of gas and that equal volumes of different gases, compared under identical conditions of temperature and pressure, contain equal numbers of molecules.

Boyle's law（波义耳定律）states that the volume of a fixed amount of gas at a constant temperature is inversely proportional to the gas pressure.

Charles's law（查理定律）states that the volume of a fixed amount of gas at a constant pressure is directly proportional to the Kelvin (absolute) temperature.

Dalton's law of partial pressures（道尔顿分压定律）states that in a mixture of gases, the total pressure is the sum of the partial pressures of the gases present.

Diffusion（扩散）refers to the spreading of a substance (usually a gas or liquid) into a region where it is not originally present as a result of random molecular motion.

Effusion（逸出）is the escape of a gas through a tiny hole in its container.

An **equation of state**（状态方程）is a mathematical expression relating the amount, volume, temperature, and pressure of a substance (usually applied to gases).

The **gas constant**（气体常量），R, is the numerical constant appearing in the ideal gas equation $PV = nRT$ and in several other equations as well.

Gay-Lussac's law（盖-吕萨克定律）states that, for a sample of gas with a constant volume, its pressure, P, is directly proportional to its Kelvin temperature, T, $P_1/T_1 = P_2/T_2$.

Graham's law（格拉罕姆定律）states that the rates of effusion or diffusion of two different gases are inversely proportional to the square roots of their molar masses.

Henry's law（亨利定律）relates the solubility of a gas to the gas pressure maintained above a solution of the gaseous solute. The solubility is directly proportional to the pressure of the gas above the solution.

An **ideal (perfect) gas**（理想气体）is one whose behavior can be predicted by the ideal gas equation.

The **ideal gas equation**（理想气体方程）relates the pressure, volume, temperature, and number of moles of ideal gas (n) through the expression $PV = nRT$.

The **kinetic-molecular theory of gases**（气体分子运动理论）is a model for describing gas behavior. It is based on a set of assumptions and yields equations from which various properties of gases can be deduced.

A **partial pressure**（分压）is the pressure exerted by an individual gas in a mixture, independently of other gases. Each gas in the mixture expands to fill the container and exerts its own partial pressure.

Standard conditions of temperature and pressure (STP)（标准状况的温度及压力）refers to a gas maintained at a temperature of exactly (273.15 K) and 760 mmHg (1 atm).

The **Van der Waals equation**（范德瓦耳斯方程）is an equation of state for nonideal gases. It includes correction terms to account for intermolecular forces of attraction and for the volume occupied by the gas molecules themselves.

(6) Thermochemistry

Bomb calorimeter（弹式热量计）is a device used to measure the heat of a combustion reaction. The quantity measured is the heat of reaction at constant volume.

The **calorie (cal)**（卡路里；卡）is the quantity of heat required to change the temperature of one gram of water by one degree Celsius.

Calorimeter（热量计）is a device (of which there are numerous types) used to measure a quantity of heat.

Chemical energy（化学能）is the energy associated with chemical bonds and intermolecular forces.

A **closed system**（封闭系统）is one that can exchange energy but not matter with its surroundings.

An **endothermic reaction**（吸热反应）results in a lowering of the temperature of an isolated system or the absorption of heat by a system that interacts with its surroundings.

Energy（能量）is the capacity to do work.

Enthalpy（焓），H, is a thermodynamic function used to describe constant pressure processes.

Enthalpy change（焓变），ΔH, is the difference in enthalpy between two states of a system.

An **enthalpy diagram**（焓图）is a diagrammatic representation of the enthalpy changes in a process.

An **exothermic reaction**（放热反应）produces an increase in temperature in an isolated system or, for a system that interacts with its surroundings, the evolution of heat.

First law of thermodynamics（热力学第一定律）is an alternative of the law of conservation of energy.

A **function of state** (**state function**)（状态函数）is a property that assumes a unique value when the state or present condition of a system is defined.

Heat capacity（热容）is the quantity of heat required to change the temperature of an object or substance by one degree, usually expressed as J/℃ or cal/℃.

A **heat of reaction**（反应热）is energy converted from chemical to thermal (or vice versa) in a reaction. In an isolated system, this energy conversion causes a temperature change, and in a system that interacts with its surroundings, heat is either evolved to or absorbed from the surroundings.

Hess's law（赫斯定律）states that the enthalpy change for an overall or net process is the sum of enthalpy changes for individual steps in the processes.

The **internal energy**（内能）of a system is the total energy attributed to the particles of matter and their interactions within a system.

An **isolated system**（孤立系统）is one that exchange neither energy nor matter with its surroundings.

Kinetic energy（K. E.，动能）is energy of motion. The kinetic energy of an object with mass m and velocity v is K. E. $= \frac{1}{2}mv^2$.

The **law of conservation of energy**（能量守恒定律）states that energy can neither be created nor destroyed in ordinary processes.

An **open system**（敞开系统）is one that can exchange both matter and energy with its surroundings.

Potential energy（势能）is energy due to position or arrangement. It is the energy associated with forces of attraction and repulsion between objects.

Pressure-volume work（压力-体积功）is work associated with the expansion or compression of gases.

The **specific heat**（比热）of a substance is the quantity of heat required to change the temperature of one gram of the substance by one degree Celsius.

Specific heat capacity（比热容）is the heat capacity per gram of substance, i. e., J/(℃ • g) and **molar heat capacity**（摩尔比热容）is the heat capacity per mole, i. e., J/(℃ • mol).

The **standard state**（标准状态）of a substance refers to that substance when it is maintained at 1 bar (1 bar = 0.1 MPa) pressure and at the temperature of interest. For a gas it is the (hypothetical) pure gas behaving as an ideal gas at 1 bar pressure and the temperature of interest.

The **standard enthalpy of formation**（标准生成焓）of a substance is the enthalpy of 1 mol of the substance in its standard state from the reference forms of its elements in their standard states. The reference forms of the elements are their most stable forms at the given temperature and 1 bar pressure.

The **standard enthalpy of reaction**（标准反应焓）is the enthalpy change of a reaction in which all reactants and products are in their standard state.

The **surroundings**（环境）represent that portion of the universe with which a system interacts.

A **system**（系统）is the portion of the universe selected for a thermodynamic study.

Thermal energy（热能）is energy associated with random molecular motion.

The **thermite reaction**（铝热反应）is an oxidation reduction reaction that uses powdered aluminum metal as a reducing agent to reduce a metal oxide, such as Fe_2O_3, to the free metal.

Thermochemistry（热化学）is the study of heat associated with chemical reactions and physical processes.

Work（功）is the form of energy transfer between a system and its surroundings that can be expressed as a force acting through a distance.

(7) Entropy and Free Energy

Coupled reactions（耦合反应）are sets of chemical reactions that occur together. One (or more) of the reactions taken alone is (are) nonspontaneous and other (s), spontaneous. The overall reaction is spontaneous.

Entropy（熵）, S, is a thermodynamic property related to the number of energy levels among which the energy of a system is spread. The greater the number of energy levels for a given total energy, the greater the entropy.

Entropy change（熵变）, ΔS, is the difference in entropy between two states of a system.

Gibbs energy（吉布斯自由能）, G, is a thermodynamic function designed to produce a criterion for spontaneous change. It is defined through the equation $G = H - TS$.

Gibbs energy change（吉布斯自由能变）, ΔG, is the change in Gibbs energy that accompanies a process and can be used to indicate the direction of spontaneous change. For a spontaneous process at constant temperature and pressure, $\Delta G < 0$ kJ/mol.

A **nonspontaneous process**（非自发过程）is one that will not occur naturally. A nonspontaneous process can be brought about only by intervention from outside the system, as in the use of electricity to decompose a chemical compound (electrolysis).

A **reversible process**（可逆过程）is one that can be made to reverse direction by just an infinitesimal change in a system property.

The **second law of thermodynamics**（热力学第二定律）relates to the direction of spontaneous change. One statement of the law is that all spontaneous processes produce an increase in the entropy of the universe.

The **standard molar entropy** ($S^{\ominus}$)（标准摩尔熵）is the absolute entropy evaluated when one mole of a substance is in its standard state at a particular temperature.

Standard Gibbs energy change（标准吉布斯自由能变），$\Delta G^{\ominus}$ is the Gibbs energy change of a process when the reactants and products are all in their standard states. The equation relating standard free energy change to the equilibrium constant is $\Delta G^{\ominus} = -RT \ln K$.

The **standard Gibbs energy of formation**（标准生成吉布斯自由能），$\Delta G_f^{\ominus}$ is the standard free energy change associated with the formation of 1 mol of compound from its elements in their most stable forms at 1 bar pressure.

A **spontaneous (natural) process**（自发过程）is one that is able to take place in a system left to itself. No external action is required to make the process go, although in some cases the process may take a very long time.

Thermodynamics（热力学）is the scientific study of the interconversion of heat and other kinds of energy.

The **third law of thermodynamics**（热力学第三定律）states that the entropy of a pure perfect crystal is zero at the absolute zero of temperature, 0 K.

(8) Chemical Equilibrium

Activity（活度）is the effective concentration of a species. It is obtained as the product of an activity coefficient and the ratio of the stoichiometric concentration or pressure to that of a reference state.

Equilibrium（平衡）refers to a condition where forward and reverse processes proceed at equal rates and no further net change occurs. For example, amounts of reactants and products in a reversible reaction remain constant over time.

The **equilibrium constant**（平衡常数）is the numerical value of the equilibrium constant expression.

An **equilibrium constant expression**（平衡常数表达式）describes the relationship among the concentrations (or partial pressures) of the substances present in a system at equilibrium.

Fractional precipitation（分步沉淀）is a technique in which two or more ions in solution, each capable of being precipitated by the same reagent, are separated by the use of that reagent.

The **formation constant**（形成常数），K_f, describes equilibrium among a complex ion, the free metal ion, and ligands.

An **ion pair**（离子对）is an association of a cation and an anion in solution. Such combinations, when they occur, can have a significant effect on solution equilibria.

An **ion product**（离子积）, Q_{sp}, is formulated in the same manner as a solubility product constant, K_{sp}, but with nonequilibrium concentration terms. A comparison of Q_{sp} and K_{sp} provides a criterion for precipitation from solution.

An **irreversible process**（不可逆过程）takes place in one or several finite steps such that the system is not in equilibrium with its surroundings.

K_c is the relationship among the concentrations of the reactants and products in a reversible reaction at equilibrium. Concentrations are expressed as molarities.

K_p, the partial pressure equilibrium constant, is the relationship that exists among the partial pressures of gaseous reactants and products in a reversible reaction at equilibrium. Partial pressures are expressed in atm.

Le Châtelier's principle（勒夏特列原理）states that an action that tends to change the temperature, pressure, or concentrations of reactants in a system at equilibrium stimulates a response that partially offsets the change while a new equilibrium condition is established.

Qualitative cation analysis（定性阳离子分析）is a laboratory method, based on a variety of solution equilibrium concepts, for determining the presence or absence of certain cations in a sample.

The **reaction quotient**（反应商）, Q, is a ratio of concentration terms (or partial pressures) having the same form as an equilibrium constant expression, but usually applied to nonequilibrium conditions. In a rearrangement reaction, a molecule is converted into another of its isomeric forms.

A **reversible process**（可逆过程）is one that can be made to reverse direction by just an infinitesimal change in a system property.

The **salt effect**（盐效应）is that of ions different from those directly involved in a solution equilibrium. The salt effect is also known as the diverse or "uncommon" ion effect.

(9) Solutions

An **alloy**（合金）is a mixture of two or more metals. Some alloys are solid solutions, some are heterogeneous mixtures, and some are intermetallic compounds.

Amalgams（汞合金）are metal alloys containing mercury. Depending on their compositions, some are liquid and some are solid.

An **azeotrope**（共沸物）is a solution that boils at a constant temperature, producing vapor of the same composition as the liquid. In some cases, the azeotrope boils at a lower temperature than the solution components, in other cases, at a higher temperature.

A **colloid**（胶体）is a mixture that contains particles that are larger than ions or molecules but are still submicroscopic.

Colligative properties（依数性质）(e.g., vapor pressure lowering, freezing point depression, boiling point elevation, osmotic pressure) have values that depend on the number of solute particles in a solution but not on their identity.

An **electrolyte**（电解质）is a substance that provides ions when dissolved in water.

An **ideal solution**（理想溶液）has $\Delta H_{soln} = 0$ kJ/mol and certain properties (notably vapor pressure) that are predictable from the properties of the solution components.

Molality (*m*)（质量摩尔浓度）is a solution concentration expressed as the amount of solute, in moles, divided by the mass of solvent, in kg.

Mole fraction（摩尔分数）describes a mixture in terms of the fraction of all the molecules that are of a particular type. It is the amount of one component, in moles, divided by the total amount of all the substances in the mixture.

A **mole percent**（摩尔百分数）is a mole fraction expressed on a percentage basis, that is, mole fraction × 100%.

Nonelectrolyte（非电解质）is a substance that is essentially non-ionized, both in the pure state and in solution.

Osmosis（渗透）is the net flow of solvent molecules through a semipermeable membrane, from a more dilute solution (or from the pure solvent) into a more concentrated solution.

Osmotic pressure（渗透压）is the pressure that would have to be applied to a solution to stop the passage through a semipermeable membrane of solvent molecules from the pure solvent.

The term **ppb**（parts per billion）（十亿分之一）refers to the number of parts of a component to one billion parts of the medium in which it is found.

The term **ppm**（parts per million）（百万分之一）refers to the number of parts of a component to one million parts of the medium in which it is found.

The term **ppt**（parts per trillion）（万亿分之一）refers to the number of parts of a component to one trillion parts of the medium in which it is found.

Raoult's law（拉乌尔定律）states that the vapor pressure of a solution component is equal to the product of the vapor pressure of the pure liquid and its mole fraction in solution: $P_A = \chi_A P^{\ominus}$.

Recrystallization（重结晶）is a method of purifying a substance by crystallizing the pure solid from a saturated solution while impurities remain in solution.

A **saturated solution**（饱和溶液）is one that contains the maximum quantity of solute that is normally possible at the given temperature.

The **solubility**（溶解度）of a substance is the concentration of its saturated solution.

A **solute**（溶质）is a solution component that is dissolved in a solvent. A solution may have several solutes, with the solutes generally present in lesser amounts than is the solvent.

The **solvent**（溶剂）is the solution component in which one or more solutes are dissolved. Usually the solvent is present in greater amount than the solutes and determines the state of matter in which the solution exists.

A **strong electrolyte**（强电解质）is a substance that is completely ionized in solution.

A **supersaturated**（过饱和）solution contains more solute than normally expected for a saturated solution, usually prepared from a solution that is saturated at one temperature by changing its temperature to one where supersaturation can occur.

An **unsaturated solution**（不饱和溶液）contains less solute than the solvent is capable of dissolving under the given conditions.

Reverse osmosis（逆渗透）is the passage through a semipermeable membrane of solvent molecules from a solution into a pure solvent. It can be achieved by applying to the solution a pressure in excess of its osmotic pressure.

A **weak electrolyte**（弱电解质）is a substance that is only partially ionized in solution in a reversible reaction.

(10) Acids and Bases

An *acid*（酸）is (1) a hydrogen-containing compound that can produce hydrogen ions, (Arrhenius theory); (2) a proton donor (Brønsted Lowry theory); (3) an atom, ion, or molecule that can accept a pair of electrons to form a covalent bond (Lewis theory).

An **acid-base indicator**（酸碱指示剂）is a substance used to measure the pH of a solution or to signal the equivalence point in an acid base titration. The nonionized weak acid form has one color and the anionic form, a different color.

An **acid ionization constant**（酸电离常数）, K_a, is the equilibrium constant for the ionization reaction of a weak acid.

Amphiprotic（两性的）substances can act either as an acid or as a base.

An **adduct**（加合物）is a compound formed by joining together two simpler molecules through a coordinate covalent bond.

A **base**（碱）is (1) a compound that produces hydroxide ions, in water solution (Arrhenius theory); (2) a proton acceptor (Brønsted-Lowry theory); (3) an atom, ion, or molecule that can donate a pair of electrons to form a covalent bond (Lewis theory).

A **base ionization constant**（碱电离常数）, K_b, is the equilibrium constant for the ionization reaction of a weak base.

Basicity（碱度）is a measure of the tendency of an electron pair donor to react with a proton.

Buffer capacity（缓冲容量）refers to the amount of acid and/or base that a buffer solution can neutralize while maintaining an essentially constant pH.

Buffer range（缓冲范围）is the range of pH values over which a buffer solution can maintain a fairly constant pH.

A **buffer solution**（缓冲溶液）resists a change in its pH. It contains components capable of neutralizing small added amounts of acids and base.

The **common-ion effect**（同离子效应）describes the effect on an equilibrium by a substance that furnishes ions that can participate in the equilibrium.

A **conjugate acid**（共轭酸）is formed when a Brønsted-Lowry base gains a proton. Every base has a conjugate acid.

A **conjugate acid-base pair**（共轭酸碱对）is pair of molecules or ions for which the chemical formulas differ by a single proton.

A **conjugate base**（共轭碱）remains after a Brønsted-Lowry acid has lost a proton. Every acid has a conjugate base.

Degree of ionization（电离度）refers to the extent to which molecules of a weak acid or weak base ionize. The degree of ionization increases as the weak electrolyte solution is diluted.

The **end point**（终点）is the point in a titration where the indicator used changes color.

A properly chosen indicator has its end point coming as closely as possible to the equivalence point of the titration.

The **equivalence point**（当量点）of a titration is the condition in which the reactants are in stoichiometric proportions. They consume each other, and neither reactant is in excess.

Hydrolysis（水解）is a special name given to acid-base reactions in which ions act as acids or bases. As a result of hydrolysis, many salt solutions are not pH neutral, that is, pH ≠7.

The **Henderson-Hasselbalch equation**（亨德森-哈塞尔巴尔赫方程）has the form, pH=pK_a+lg[conjugate base]/[acid], in which stoichiometric concentrations of the weak acid and its conjugate base are used in place of the equilibrium concentrations. There are limitations on its validity.

Hydronium ion（水合氢离子）, H_3O^+, is the form in which protons are found in aqueous solution. The terms "hydrogen ion" and "hydronium ion" are often used synonymously.

The **ion product of water**（水的离子积）, K_w, is the product of $[H_3O^+]$ and $[OH^-]$ in pure water or in an aqueous solution. This product has a unique value that depends only on temperature. At 25℃, $K_w=1.0\times10^{-14}$.

The **percent ionization**（电离度）of a weak acid or a weak base is the percent of its molecules that ionize in an aqueous solution.

pH is a shorthand designation for $[H_3O^+]$ in a solution. It is defined as pH=$-lg[H_3O^+]$.

pOH is a shorthand designation for $[OH^-]$ in a solution: pOH=$-lg[OH^-]$

A **polyprotic acid**（多质子酸）is capable of losing more than a single proton per molecule in acid base reactions. Protons are lost in a stepwise fashion, with the first proton being the most readily lost.

A **proton acceptor**（质子受体）is a base in the Brønsted-Lowry acid-base theory.

A **proton donor**（质子供体）is an acid in the Brønsted-Lowry acid-base theory.

Salts（盐）are ionic compounds in which hydrogen atoms of acids are replaced by metal ions. Salts are produced by the neutralization of acids with bases.

Self-ionization（自电离）is an acid-base reaction in which one molecule acts as an acid and donates a proton to another molecule of the same kind acting as a base.

Standardization of a solution（溶液标定）refers to establishing the exact concentration of the solution, usually through a titration.

A **strong acid**（强酸）is an acid that is completely ionized in aqueous solution.

A **strong base**（强碱）is a base that is completely ionized in aqueous solution.

A **weak acid**（弱酸）is an acid that is only partially ionized in aqueous solution in a reversible reaction.

A **weak base**（弱碱）is a base that it only partially ionized in aqueous solution in a reversible reaction.

The **titrant**（滴定剂）is the solution that is added in a controlled fashion through a burette in a titration reaction.

Titration（滴定）is a procedure for carrying out a chemical reaction between two solutions by the controlled addition (from a burette) of one solution to the other. In a titration a means must be found, as by the use of an indicator, to locate the equivalence point.

A **titration curve**（滴定曲线）is a graph of solution pH versus volume of titrant. It outlines how pH changes during an acid-base titration, and it can be used to establish such features as the equivalence point of the titration.

(11) Chemical Kinetics

An **activated complex**（活化络合物）is an intermediate in a chemical reaction formed through collisions between energetic molecules. Once formed, it dissociates either into the products or back to the reactants.

Activation energy（活化能）is the minimum total kinetic energy that molecules must bring to their collisions for a chemical reaction to occur.

Active sites（活性位点）are the locations at which catalysis occurs, whether on the surface of a heterogeneous catalyst or an enzyme.

A **bimolecular process**（双分子过程）is an elementary process involving the collision of two molecules and consumed in a subsequent one. As a result, the species does not appear in the equation for the overall reaction.

A **catalyst**（催化剂）provides an alternative mechanism of lower activation energy for a chemical reaction. The reaction is speeded up, and the catalyst is regenerated.

An **enzyme**（酶）is a high molar mass protein that catalyzes biological reactions.

A **first-order reaction**（一级反应）is one for which the sum of the concentration-term exponents in the rate equation is 1.

The **half-life**（半衰期）($t_{1/2}$) of a reaction is the time required for one-half of a reactant to be consumed. In a nuclear decay process, it is the time required for one-half of the atoms present in a sample to undergo radioactive decay.

Heterogeneous catalysis（多相催化剂）refers to a catalytic reaction taking place in different phases.

Homogeneous catalysis（均相催化剂）refers to a catalytic reaction taking place in a single phase.

The **initial rate of a reaction**（反应的初始速率）is the rate of a reaction immediately after the reactants are brought together.

An **instantaneous rate of reaction**（反应的瞬时速率）is the exact rate of a reaction at some precise point in the reaction. It is obtained from the slope of a tangent line to a concentration-time graph.

An **integrated rate law**（equation）（积分的速率方程）is derived from a rate law (equation) by the calculus technique of integration. It relates the concentration of a reactant (or product) to elapsed time from the start of a reaction. The equation has different forms depending on the order of the reaction.

The **order of a reaction**（反应级数）relates to the exponents of the concentration terms in the rate law for a chemical reaction. The order can be stated with respect to a particular reactant (first order in A, second order in B⋯) or, more commonly, as the overall order.

The overall order is the sum of the concentration-term exponents.

The **rate constant**（速率常数）, k, is the proportionality constant in a rate law that permits the rate of a reaction to be related to the concentrations of the reactants.

A **rate-determining step**（速控步骤） in a reaction mechanism is an elementary process that is instrumental in establishing the rate of the overall reaction, usually because it is the slowest step in the mechanism.

The **rate law**（**rate equation**）（速率定律，速率方程） for a reaction relates the reaction rate to the concentrations of the reactants. It has the form: rate $= k[A]^m[B]^n \cdots$.

A **reaction intermediate**（反应中间体） is a species formed in one elementary reaction in a reaction mechanism.

A **reaction mechanism**（反应机理） is a set of elementary steps or processes by which a reaction is proposed to occur. The mechanism must be consistent with the stoichiometry and rate law of the overall reaction.

The **rate of reaction**（反应速率） describes how fast reactants are consumed and products are formed, usually expressed as change of concentration per unit time.

A **reaction profile**（反应图） is a graphical representation of a chemical reaction in terms of the energies of the reactants, activated complex(es), and products.

A **second-order reaction**（二级反应） is one for which the sum of the concentration-term exponents in the rate equation is 2.

A **substrate**（底物） is the substance that is acted upon by an enzyme in an enzyme-catalyzed reaction. The substrate is converted to products, and the enzyme is regenerated.

The **transition state**（过渡状态） in a chemical reaction is an intermediate state between the reactants and products.

A **unimolecular process**（单分子过程） is an elementary process in a reaction mechanism in which a single molecule, when sufficiently energetic, dissociates.

A **zero-order reaction**（零级反应） proceeds at a rate that is independent of reactant concentrations. The sum of the concentration-term exponent(s) in the rate equation is equal to zero.

(12) Atomic Structures and Properties

An **angular wave function**（角度波函数） is the part of a wave function that depends on the angles θ and ϕ when the Schrödinger wave equation is expressed in spherical polar coordinates.

Atomic (line) spectra（原子光谱） are produced by dispersing light emitted by excited gaseous atoms. Only a discrete set of wavelength components (seen as colored lines) is present in a line spectrum.

Cathode rays（阴极射线） are negatively charged particles (electrons) emitted at the negative electrode (cathode) in the passage of electricity through gases at very low pressures.

Degenerate orbitals（简并轨道） are orbitals that are at the same energy level.

A **diamagnetic**（反磁性的） substance has all its electrons paired and is slightly repelled by a magnetic field.

Diagonal relationships（对角线规则） refer to similarities that exist between certain pairs of elements in different groups and periods of the periodic table, such as Li and Mg, Be and Al, and B and Si.

Effective nuclear charge（有效核电荷）is the positive charge acting on a particular electron in an atom. Its value is the charge on the nucleus, reduced to the extent that other electrons screen the particular electron from the nucleus.

An **electron configuration**（电子构型）is a designation of how electrons are distributed among various orbitals in an atom.

Electromagnetic radiation（电磁辐射）is a form of energy propagated as mutually perpendicular electric and magnetic fields. It includes visible light, infrared, ultraviolet, X ray, and radio waves.

An **electron affinity**（**EA**）（电子亲和能）is the energy released (the negative of the enthalpy change, ΔH) when an atom in the gas phase accepts an electron.

Electronegativity（**EN**）（电负性）is a measure of the electron-attracting power of a bonded atom; metals have low electronegativities, and nonmetals have high electro-negativities.

The **electronegativity difference**（电负性差）between two atoms that are bonded together is used to assess the degree of polarity in the bond.

Electron spin（电子自旋）is a characteristic of electrons giving rise to the magnetic properties of atoms. The two possibilities for electron spin are $+1/2$ and $1/2$.

In an **excited state**（激发态）of an atom, one or more electrons are promoted to a higher energy level than in the ground state.

The **frequency**（频率）of a wave motion is the number of wave crests or troughs that pass through a given point in a unit of time^{-1}. It is expressed by the unit s^{-1}, which is also called a hertz, Hz.

The **ground state**（基态）is the lowest energy state for the electrons in an atom or molecule.

The **Heisenberg uncertainty principle**（海森堡不确定性原理）states that, when measuring the position and momentum of fundamental particles of matter, uncertainties in measurement are inevitable.

Hund's rule（**rule of maximum multiplicity**）（洪特规则）states that whenever orbitals of equal energy are available, electrons occupy these orbitals singly before any pairing of electrons occurs.

An **ionic radius**（离子半径）is the radius of a cation or an anion.

Ionization energy（**IE**）（电离能）is the minimum energy required to remove an electron from an atom in the gas phase.

Isoelectronic（等电子的）species have the same number of electrons (usually in the same configuration). Ne is isoelectronic, as are CO and N_2.

The **lanthanide contraction**（镧系收缩）refers to the decrease in atomic size in a series of elements in which an subshell fills with electrons (an inner transition series). It results from the ineffectiveness of electrons in shielding outer-shell electrons from the nuclear charge of an atom.

A **lone pair**（孤对电子）is a pair of electrons found in the valence shell of an atom and not involved in bond formation.

Metallic radius（金属半径）is one-half the distance between the centers of adjacent atoms in a solid metal.

An **orbital**（轨道）is a mathematical function used to describe regions in an atom where the electron density of the possibility of finding an electron is high. The several kinds of orbitals (s, p, d, f···) differ from one another in the shapes of the regions of high electron charge density they describe.

An **orbital diagram**（轨道图）is a representation of an electron configuration in which the most possible orbital designation and spin of each electron in the atom are indicated.

The **Pauli exclusion principle**（泡利不相容原理）states that no two electrons may have all four quantum numbers alike. This limits occupancy of an orbital to two electrons with opposing spins.

A **paramagnetic**（顺磁的）substance has one or more unpaired electrons in its atoms or molecules. It is attracted into a magnetic field.

A **period**（周期）is a horizontal row of the periodic table. All members of a period have atoms with the same highest principal quantum number.

The **periodic law**（周期律）refers to the periodic recurrence of certain physical and chemical properties when the elements are considered in terms of increasing atomic number.

The **periodic table**（周期表）is an arrangement of the elements, by atomic number, in which elements with similar physical and chemical properties are grouped together in vertical columns.

The **photoelectric effect**（光电效应）is the emission of electrons by certain materials when their surfaces are struck by electromagnetic radiation of the appropriate frequency.

A **photon**（光子）is a particle of light. The energy of a beam of light is concentrated into these photons.

A **principal electronic shell** (**level**)（主电子层）refers to the collection of all orbitals having the same value of the principal quantum number, n. For example, the $3s$, $3p$, and $3d$ orbitals comprise the third principal shell ($n=3$).

A **quantum**（量子）refers to a discrete unit of energy that is the smallest quantity by which the energy of a system can change.

Quantum numbers（量子数）are integral numbers whose values must be specified in order to solve the equations of wave mechanics. Three different quantum numbers are required: the principal quantum number, n; the orbital angular momentum quantum number, l; and the magnetic quantum number, m_l. The permitted values of these numbers are interrelated.

A **radial wave function**（径向波函数）is the part of a wave function that depends only on the distance r when the Schrödinger wave equation is expressed in spherical polar coordinates.

The **Schrödinger equation**（薛定谔方程）describes the electron in a hydrogen atom as a matter wave. Solutions to the Schrödinger equation are called wave functions.

The **shielding effect**（屏蔽效应）refers to the effect of inner-shell electrons in shielding or screening outer-shell electrons from the full effects of the nuclear charge. In effect the inner electrons partially reduce the nuclear charge.

$spdf$ **notation**（$spdf$ 符号）is a method of describing electron configurations in which the numbers of electrons assigned to each orbital are denoted as superscripts. An **electron configuration** is a designation of how electrons are distributed among various orbitals in an atom.

A **subshell**（亚层）refers to a collection of orbitals of the same type. For example, the three $2p$ orbitals constitute the $2p$ subshell.

Valence electrons（价电子层）are electrons in the electronic shell of highest principal quantum number, that is, electrons in the outermost shell.

Wave-particle duality（波粒二象性）was postulated by de Broglie and states that at times particles of matter have wave-like properties and vice versa. This was demonstrated in the diffraction pattern observed in the diffraction pattern observed when electrons were directed at a nickel crystal.

(13) Chemical Bonding

An **antibonding molecular orbital**（反键轨道）describes regions in a molecule in which there is a low electron probability or charge density between two bonded atoms.

An **average bond energy**（平均键能）is the average of bond-dissociation energies for a number of different species containing a particular covalent bond.

Band theory（价带理论）is a form of molecular orbital theory to describe bonding in metals and semiconductors.

Bond dissociation energy（键解离能）is the quantity of energy required to break one mole of covalent bonds in a gaseous species.

A **bonding molecular orbital**（成键分子轨道）describes regions of high electron probability or charge density in the internuclear region between two bonded atoms.

Bond length（**bond distance**）（键长）is the distance between the centers of two atoms joined by a covalent bond.

A **bond angle**（键角）is the angle between two covalent bonds. It is the angle between hypothetical lines joining the number of third atom to which they are covalently bonded.

Bond order（键级）is one-half the difference between the number of electrons in bonding and antibonding molecular orbitals in a covalent bond. A single bond has a bond order of 1; a double bond, 2; and a triple bond, 3.

A **bond pair**（键对）is a pair of electrons involved in covalent bond formation.

A **covalent bond**（共价键）is formed when electrons are shared between a pair of atoms. In valence bond theory, the sharing of the electrons is said to occur in the region in which atomic orbitals overlap.

Covalent radius（共价半径）is one-half the distance between the centers of two atoms that are bonded covalently. It is the atomic radius associated with an element in its covalent compound.

A **delocalized molecular orbital**（离域分子轨道）describes a region of high electron probability or charge density that extends over three or more atoms.

Dipole moment（偶极矩）is a measure of the extent to which a separation exists between the centers of positive and negative charge within a molecule.

Electron-group geometry（电子几何构型）refers to the geometrical distribution about a central atom of the electron pairs in its valence shell.

Formal charge（形式电荷）is the number of outer shell (valence) electrons in an isolated atom minus the number of electrons assigned to that atom in a Lewis structure.

Hybridization（杂化）refers to combining pure atomic orbitals to generate hybrid orbitals in the valence bond approach to covalent bonding.

A **hybrid orbital**（杂化轨道）is one of a set of identical orbitals reformulated from pure atomic orbitals and used to describe certain covalent bonds.

Lewis structure（路易斯结构式）is a combination of Lewis symbols that depicts the transfer or sharing of electrons in a chemical bond.

In the **Lewis symbol**（路易斯符号）of an element, valence electrons are represented by dots placed around the chemical symbol of the element.

The **Lewis theory**（路易斯理论）refers to a description of chemical bonding through Lewis symbols and Lewis structures in accordance with a particular set of rules.

Molecular geometry（分子几何构型）refers to the geometric shape of a molecule of polyatomic ion. In a species in which all electron pairs are bond pairs, the molecular geometry is the same as the electron-group geometry. In other cases, the two properties are related but not the same.

Molecular orbital theory（分子轨道理论）describes the covalent bonds in a molecule by replacing atomic orbitals of the component atoms by molecule as a whole. A set of rules is used to assign electrons to these molecular orbitals, thereby yielding the electronic structure of the molecule.

An **octet**（八隅体）refers to eight electrons in the outermost (valence) electronic shell of an atom in a Lewis structure.

The **octet rule**（八隅体规则）states that the number of electrons associated with bond pairs and lone pairs of electrons for each of the Lewis structure (except H) will be eight.

A **pi bond**（π 键）results from the end-to-end overlap of p orbitals, producing a high electron charge density above and below the line joining the bonded atoms.

In a **polar covalent bond**（极性共价键）a separation exists between the centers of positive and negative charge in the bond.

In a **polar molecule**（极性分子）, the presence of one or more polar covalent bonds leads to a separation of the positive and negative charge centers for the molecule as a whole. A polar molecule has a resultant dipole moment.

Resonance（共振）occurs when two or more plausible Lewis structure can be written for a species. The true structure is a composite or hybrid of these different contributing structures.

A **sigma bond**（σ 键）results from the end-to-end overlap of simple or hybridized atomic orbitals along the straight line joining the nuclei of the bonded atoms.

The **valence bond theory**（价键理论）treats a covalent bond in terms of the overlap of pure or hybridized atomic orbitals. Electron probability (or electron charge density) is concentrated in the region of overlap.

The **valence-shell electron-pair repulsion (VSEPR) theory**（价层电子对互斥理论）is a theory used to predict probable shapes of molecules and polyatomic ions based on the mutual repulsions of electron pairs found in the valence shell of the central atom in the structure.

(14) Liquid, Solids, and Intermolecular Forces

Adhesive forces（附着力）are intermolecular forces between unlike molecules, such as molecular of a liquid and a surface with which it is in contact.

A **body-centered cubic (bcc)**（体心立方）crystal structure is one in which the unit cell has structural units at each corner and one in the center of the cube.

Boiling（沸腾）is a process in which vaporization occurs throughout a liquid. It occurs when the vapor pressure of a liquid is equal to barometric pressure.

Cohesive forces（内聚力）are intermolecular forces between like molecules, such as within a drop of liquid.

The **critical point**（临界点）refers to the temperature and pressure at which a liquid and its vapor become identical. It is the highest temperature point on the vapor pressure curve.

Cubic closest packing（立方密堆积）is one of the two ways in which spheres can be packed to minimize the amount of the free space or voids among them.

Deposition（凝华）is the passage of molecules from the gaseous to the solid state.

Dipole-dipole interactions（偶极-偶极作用力）are attractive forces that act between polar molecules.

Dispersion (London) forces（分散力）are intermolecular forces associated with instantaneous and induced dipoles.

A **face-centered cubic (fcc)**（面心立方）crystal structure is one in which the unit cell has structural units at the eight corners and in the center of each face of the unit cell.

Freezing（凝固）is the conversion of a liquid to a solid that occurs at a fixed temperature known as the freezing point.

Hexagonal closest packing（六方密堆积）is one of the two ways in which spheres can be packed to minimize the amount of the free space or voids among them. The crystal structure based on this type of packing is referred to as **hcp**.

A **hydrogen bond**（氢键）is an intermolecular force of attraction in which an H atom covalently bonded to one atom is attracted simultaneously to another highly nonmetallic atom of the same or a nearby molecule.

An **ionic bond**（离子键）results from the transfer of electrons between metal and nonmetal atoms. Positive and negative ions are formed and held together by electrostatic attractions.

An **ionic compound**（离子化合物）is a compound consisting of positive and negative ions that are held together by electrostatic forces of attraction.

Lattice energy（晶格能）is the quantity of energy released in the formation of one mole of a crystalline ionic solid from its separated gaseous ion.

Melting(熔化) is the transition of a solid to a liquid and occurs at the melting point. The melting point and freezing point of a substance are identical.

A **phase diagram**(相图) is a graphical representation of the conditions of temperature and pressure at which solids, liquids, and gases (vapors) exist, either as single phases or states of matter or as two or more phases in equilibrium.

Polarizability(极化) describes the ease with which the electron cloud in an atom or molecule can be distorted in an electric field, that is, the ease with which a dipole can be induced.

Polymorphism(多晶型现象) refers to the existence of a solid substance in more than one crystalline form.

Sublimation(升华) is the passage of molecules from the solid to the gaseous state.

Surface tension(表面张力) is the energy or work required to extend the surface of a liquid.

A **suspension**(悬浮体) is a heterogeneous fluid containing solid particles that are sufficiently large for sedimentation and, unlike colloids, will settle.

A **triplet point**(三相点) is a condition of temperature and pressure at which three phases of a substance (usually solid, liquid, and vapor) coexist at equilibrium.

A **unit cell**(晶胞) is a small collection of atoms, ions, or molecules occupying position in a crystalline lattice. An entire crystal can be generated by straight-line displacements of the unit cell in the three perpendicular directions.

The term **van der Waals forces**(范德瓦耳斯力) is used to describe, collectively, intermolecular forces of the London type and interactions between permanent dipoles.

Vaporization(汽化) is the passage of molecules from the liquid to the gaseous state.

Vapor pressure(蒸气压) is the pressure exerted by a vapor when it is in dynamic equilibrium with its liquid at a fixed temperature.

A **vapor-pressure curve**(蒸气-压力曲线) is a graph of vapor pressure as a function of temperature.

Viscosity(黏度) refers to a liquid's resistance to flow. Its magnitude depends on intermolecular forces of attraction and in some cases, on molecular sizes and shape.

(15) Electrochemistry

The **anode**(负极；阳极) is the electrode in an electrochemical cell at which an oxidation half-reaction occurs.

A **battery**(电池) is a voltaic cell (or a group of voltaic cells connected in series (＋ to －) used to produce electricity from chemical change.

The **cathode**(正极；阴极) is the electrode of an electrochemical cell where a reduction half-reaction occurs.

Cathodic protection(阳极保护) is a method of corrosion control in which the metal to be protected is joined to a more active metal that corrodes instead. The protected metal acts as the cathode of a voltaic cell.

A **cell diagram**(电池的图示) is a symbolic representation of an electrochemical cell that indicates the substances entering into the cell reaction, electromaterials, solution concentrations, etc.

The **cell voltage (potential)** (电池电势), E_{cell}, is the potential difference (voltage) between the two electrodes of an electrochemical cell.

In a **concentration cell** (浓差电池), identical electrodes are immersed in solutions of different concentrations. The voltage (emf) of the cell is a function simply of the concentrations of the two solutions.

In a **disproportionation reaction** (歧化反应), the same substance is both oxidized and reduce.

An **electrochemical cell** (电化学电池) is a device in which the electrons transferred in an oxidation reduction reaction are made to pass through an electrical circuit.

An **electrode** (电极) is a metal surface on which an oxidation-reduction equilibrium is established between the metal and substances in solution.

Electrolysis (电解) is the decomposition of a substance, either in the molten state or in an electrolyte solution, by means of electric current.

An **electrolytic cell** (电解池) is an electrochemical cell in which a nonspontaneous reaction is carried out by electrolysis.

Electromotive force (emf) (电动势) is the potential difference between two electrodes in a voltaic cell, expressed in volts.

The **Faraday constant**, F, (法拉第常量) is the charge associated with one mole of electrons, 96485 C/mol.

A **flow battery** (流动电池) is a battery in which materials (reactants, products, electrolytes) pass continuously through the battery. The battery is simply a converter of chemical to electrical energy.

A **fuel cell** (燃料电池) is a voltaic cell in which the cell reaction is the equivalent of the combustion of a fuel. Chemical energy of the fuel is converted to electricity.

A **half-cell** (半电池) is a combination of an electrode and a solution. An oxidation-reduction equilibrium is established on the electrode. An electrochemical cell is a combination of two half-cells.

The **Nernst equation** (能斯特方程) is used to relate E_{cell}, $E_{cell}^{\ominus}$ and the activities of the reactants and products in a cell reaction.

An **overall reaction** or **overall equation** (总反应) is the overall or net change that occurs when a process is carried out in more than one step.

An **overpotential** (过电势) is the voltage in excess of the theoretical value required to produce a particular electrode reaction in electrolysis.

Oxidation (氧化) is a process in which electrons are lost and the oxidation state of some atom increases.

In an **oxidation-reduction (redox) reaction** (氧化还原反应), certain atoms undergo changes in oxidation state. The substance containing atoms whose oxidation states increase is oxidized. The substance containing atoms whose oxidation states decrease is reduced.

An **oxidation state** (氧化态) relates to the number of electrons an atom loses, gains, or shares in combining with other atoms to form molecules or polyatomic ions.

An **oxidizing agent**（**oxidant**）（氧化剂）makes possible an oxidation process by itself being reduced.

A **primary battery** or primary cell（一次电池）produces electricity from a chemical reaction that cannot be reversed. As a result the battery cannot be recharged.

A **reducing agent**（**reductant**）（还原剂）makes possible a reduction process by itself becoming oxidized.

A **reduction**（还原）process is one in which electrons are gained and the oxidation state of some atom decreases.（Reduction can only occur in combination with oxidation.）

Refining（精炼）removes impurities from the metal.

A **salt bridge**（盐桥）is a device (a U-tube filled with a salt solution) used to join two half-cells in an electrochemical cell. The salt bridge permits the flow of ions between the two half-cells.

A **secondary battery**（二次电池；蓄电池）produces electricity from a reversible chemical reaction. When electricity is passed through the battery in the reverse direction the battery is recharged.

A **standard cell potential**（标准电池电势），$E_{\text{cell}}^{\ominus}$, is the voltage of an electrochemical cell in which all species are in their standard states.

A **standard electrode potential**（标准电极电势），$E^{\ominus}$, is the electric potential that develops on an electrode when the oxidized and reduced forms of some substance are in their standard states. Tabulated data are expressed in terms of the reduction process, that is, standard electrode potentials are standard reduction potentials.

The **standard hydrogen electrode**（**SHE**）（标准氢电极）is an electrode at which equilibrium is established between H_3O^+ ($a=1$) and H_2 (g, 1 bar) on an inert (Pt) surface. The standard hydrogen electrode is *arbitrarily* assigned an electrode potential of exactly 0 V.

A **volt**（**V**）（伏）is the SI unit for cell voltage. It is defined as 1 J per coulomb.

A **voltaic**（**galvanic**）**cell**（原电池）is an electrochemical cell in which a spontaneous chemical reaction produces electricity.

(16) Coordination Compounds

A **bidentate**（双齿配体）ligand attaches itself to the central atom of a complex at two points in the coordination sphere.

A **chelate**（螯合物）results from the attachment of polydentate ligands to the central atom of a complex ion. Chelates are five- or six-membered rings that include the central atom and atoms of the ligands.

A **chelating agent**（螯合剂）is a polydentate ligand. It simultaneously attaches to two or more positions in the coordination sphere of the central atom of a complex ion.

Chelation（螯合）is the process of chelate formation.

The **chelation effect**（螯合效应）refers to an exceptional stability conferred to a complex ion when polydentate ligands are present.

Chiral（手性的）refers to a molecule with a structure that is not superimposable on its mirror image.

A **complementary color**（互补色）is a secondary color that mixes with the opposite primary color on the color wheel to produce white light in additive color mixing or black in subtractive color mixing.

A **complex**（配合物）is a polyatomic cation, anion, or neutral molecule in which groups (molecules or ions) called ligands are bonded to a central metal atom or ion.

A **complex ion**（配离子）is a complex having a net electrical charge.

Coordination compounds（配合物）are neutral complexes or compounds containing complex ions.

In a **coordinate covalent bond**（配位键）, electrons shared between two atoms are contributed by just one of the atoms. As a result, the bonded atoms exhibit formal charges.

Coordination number（配位数）is the number of positions around a central atom where ligands can be attached in the formation of a complex. Applied to a crystalline solid, coordination number signifies the number of nearest neighboring atoms (or ions of opposite charge) to any given atom (or ion) in a crystal.

Crystal field theory（晶体场理论）describes bonding in complexes in terms of electrostatic attractions between ligands and the nucleus of the central metal. Particular attention is focused on the splitting of the d energy level of the central metal.

Enantiomers (**optical isomers**)（光学异构体）are molecules whose structures are non-superimposable mirror images. The molecules are optically active, that is, able to rotate the plane of polarized light.

Ferromagnetism（铁磁性）is a property that permits certain materials (notably Fe, Co, and Ni) to be made into permanent magnets. The magnetic moments of individual atoms are aligned into domains. In the presence of a magnetic field, these domains orient themselves to produce a permanent magnetic moment.

Geometric isomerism（几何异构）in complexes refers to the nonequivalent structures based on the positions at which ligands are attached to the metal center.

In a **high-spin complex**（高自旋配合物）, weak crystal field splitting leads to a maximum number of unpaired electrons in the d subshell of the central metal atom or ion.

The **inert complex**（惰性配合物）is the term used to describe a complex ion in which the exchange of ligands occurs very slowly.

Isomers（异构体）are two or more compounds having the same formula but different structures and therefore different properties.

The **labile complex**（易变配合物）is the term used to describe a complex ion in which a rapid exchange of ligands occurs.

Ligands（配体）are the groups that are coordinated (bonded) to the central atom in a complex.

In a **low-spin complex**（低自旋配合物）, strong crystal field splitting leads to a minimum number of unpaired electrons in the d subshell of the central metal atom or ion.

A **monodentate ligand**（单齿配体）is a ligand that is able to attach to a metal center in a complex at only one position and using just one lone pair of electrons.

Optical isomerism（光学异构）results from the presence of a chiral atom in a structure, leading to a pair of optical isomers that differ only in the direction that they rotate the plane of polarized light.

Pairing energy（配对能）is the energy requirement to force an electron into an orbital that is already occupied by one electron.

A **polydentate ligand**（多齿配体）is capable of donating more than a single electron pair to the metal center of a complex, from different atoms in the ligand and to different sites in the geometric structure.

A **primary color**（基本色）is one of a set of colors that when added together as light produce white light. Subtractive mixing leads to an absence of color (black). Red, yellow, and blue are a set of primary colors.

The **spectrochemical series**（光谱化学序列）is a ranking of ligand abilities to produce a splitting of the d energy level of a central metal ion in a complex ion.

Structural isomers（结构异构体）have the same number and kinds of atoms, but they differ in their structural formulas.

In **stereoisomers**（立体异构体）, the number and types of atoms and bonds in molecules are the same, but certain atoms are oriented differently in space. Cis and trans isomerism is one type of stereoisomerism; optical isomerism is another.

The term **trans**（反式）is used to describe geometric isomers in which two groups are attached on opposite sides of a double bond in an organic molecule, or at opposite corners of a square in a square-planar complex, or at positions above and below the central plane of an octahedral complex.

(17) Nuclear Chemistry

An **α (alpha) particle**（α 粒子）is a combination of two protons and two neutrons identical to the helium ion, $^4He^{2+}$. α particles are emitted in some radioactive decay processes.

A **β (beta) particle**（β 粒子）is an electron emitted as a result of the conversion of a neutron to a proton in certain atomic nuclei undergoing radioactive decay.

Control rods（控制棒）are neutron-absorbing metal rods (e.g., Cd) that are used to control the neutron flux in a nuclear and thereby control the rate of the fission reaction.

A **decay constant**（衰变常数）is a first-order rate constant describing radioactive decay.

Electron capture (E. C.)（电子捕获）is a form of radioactive decay in which an electron from an inner electronic shell is absorbed by a nucleus. In the nucleus the electron is used to convert a proton to a neutron.

γ (gamma) rays（γ 射线）are a form of electro-magnetic radiation of high penetrating power emitted by certain radioactive nuclei.

The **neutron number**（中子数）is the number of neutrons in the nucleus of an atom. It is equal to the mass number (A) minus the atomic number (Z).

Neutrons（中子）are electrically neutral fundamental particles of matter found in all atomic nuclei except that of the simple hydrogen atom, protium, ^1H.

Nuclear fission（核裂变）is a radioactive decay process in which a heavy nucleus breaks up into two lighter nuclei and several neutrons, accompanied by the release of energy.

In **nuclear fusion**（核聚变）small atomic nuclei are fused into larger ones, with some of their mass being converted to energy.

A **nuclear reaction**（核反应）is a device in which nuclear fission is carried out as a controlled chain reaction. That is, neutrons produced in one fission event trigger the fission of other nuclei, and so on.

The **half-life**（半衰期）of a reaction is the time required for one-half of a reactant to be consumed. In a nuclear decay process, it is the time required for one-half of the atoms present in a sample to undergo radioactive decay.

A **moderator**（调节器）slows down energetic neutrons from a fission process so that they are able to induce additional fission.

Nuclear binding energy（核结合能）is the energy released when nucleons (protons and neutrons) are fused into an atomic nucleus. This energy replaces an equivalent quantity of matter.

A **nuclear equation**（核方程）represents the changes that occur during a nuclear process. The target nucleus and bombarding particle are represented on the left side of the equation, and the product nucleus and ejected particle on the right side.

A **positron**（β^+）（正电子）is a positive electron emitted as a result of the conversion of a proton to a neutron in a radioactive nucleus.

A **rad**（拉德）is a quantity of radiation able to deposit 1×10^{-2} J of energy per kilogram of matter.

The **radioactive decay law**（放射性衰变定律）states that the rate of decay of a radioactive material (the activity, A) is directly proportional to the number of atoms present.

A **radioactive decay series**（放射性衰变系）is a succession of individual steps whereby an initial radioactive isotope (e.g., ^{236}U) is ultimately converted to a stable isotope (e.g., ^{206}Pb).

Radioactivity（放射性）is a phenomenon in which small particles of matter (or particles) and/or electromagnetic radiation (rays) are emitted by unstable atomic nuclei.

A **rem**（雷姆）is a unit of radiation related to the rad, but taking into account the varying effects on biological matter of different types of radiation of the same energy.

(18) Organic Chemistry

Absolute configuration（绝对构型）refers to the spatial arrangement of the groups attached to a chiral carbon atom. The two possibilities are D and L.

An **achiral**（非手性的）molecule has a structure that is superimposable on its mirror image.

The **acyl group**（酰基）is C(O)R. If R = H, this is called the formyl group; R=CH$_3$, **acetyl**; and R = C$_6$H$_5$, **benzoyl**.

In an **addition reaction**（加成反应）, functional group atoms are joined to the carbon atoms at points of unsaturation in alkene and alkyne hydrocarbon molecules.

In **addition-elimination reaction**（加成-消去反应）is the overall reaction that occurs when compounds are interconverted. It involves (1) a nucleophilic addition to the carbonyl carbon to form a tetrahedral intermediate, followed by (2) an elimination reaction that regenerates the carbonyl group.

Alcohols（醇）contain the functional group —OH and have the general formula ROH.

Aldehydes（醛）have the general formula RCHO.

Alicyclic（脂环的）hydrocarbon molecules have their carbon atom skeletons arranged in rings and resemble aliphatic (rather than aromatic) hydrocarbons.

Aliphatic（脂肪族的）hydrocarbon molecules have their carbon atom skeletons arranged in straight or branched chains.

Alkane（烷）hydrocarbon molecules have only single covalent bonds between carbon atoms. In their chain structures alkanes have the general formula C_nH_{2n+2}.

Alkene（烯）hydrocarbon have one or more carbon-carbon double bonds in their molecules. The simple alkenes have the general formula C_nH_{2n}.

Alkyne（炔）hydrocarbon have one or more carbon-carbon triple bonds in their molecules. The simple alkynes have the general formula C_nH_{2n-2}.

Alkyl groups（烷基）are alkane hydrocarbon molecules from which one hydrogen atom has been extracted. For example, the group CH_3 is the **methyl** group; CH_2CH_3 is the **ethyl** group.

An **amide**（酰胺）is derived from the ammonium salt of a carboxylic acid and has the general formula $RC(O)NH_2$.

An **amine**（胺）is an organic base having the formula RNH_2（primary）, R_2NH（secondary）, or R_3N（tertiary）depending on the number of hydrogen atoms of an NH_3 molecule that are replaced by R groups.

An **aprotic solvent**（非质子溶剂）is a solvent whose molecules do not have a hydrogen atom bonded to an electronegative element.

Aromatic（芳香族的）compounds are organic substance whose carbon atom skeletons are arranged in hexagonal rings, based on benzene, C_6H_6.

Asymmetric（不对称的）is the term used to describe a C atom with four different substituent groups. Molecule with such a C atom is chiral.

The **carbonyl group**（羰基）, —C(O), is found in aldehydes, ketones, and carboxylic acids.

The **carboxyl group**（羧基）is —COOH.

A **carboxylic acid**（羧酸）has one or more carboxyl groups attached to a hydrocarbon chain or ring structure.

In **chain reaction polymerization**（连锁聚合反应）, a reaction is initiated by "opening up" a carbon-carbon double bond. Monomer units add to free-radical intermediates to produce a long-chain polymer.

The term **cis**（顺式）describes geometric isomers in which two groups are attached on the same side of a double bond in an organic molecule, or along the same edge of a square in a square-planar complex, or at two adjacent vertices of an octahedral complex.

Chiral（手性的）refers to a molecule with a structure that is not superimposable on its mirror image.

A **condensation reaction**（缩合反应）is one in which two molecules are combined by eliminating a small molecule (such as H_2O) between them.

A **condensed structural formula**（结构简式）is a simplified representation of a structural formula.

Conformations（构型）refer to the different spatial arrangement possible in a molecule. Examples are the "boat" and "chair" forms of cyclohexane.

Diastereomers（非对映异构体）are optically active isomers of a compound, but their structures are not mirror image.

Detergents（洗涤剂）are cleansing agents that act by emulsifying oils. Most common among synthetic detergents are the salts of organic sulfonic acids, $R-SO_3^- Na^+$.

Dextrorotatory（右旋的）means the ability to rotate the plane of polarized light to the right, designated (+).

An **E1 reaction**（单分子消除反应）is an elimination reaction in which the rate-determining step is unimolecular.

An **E2 reaction**（双分子消除反应）is an elimination reaction in which the rate-determining step is bimolecular.

An **electrophile**（亲电试剂）contains an electron attracting region of positive charge (an electrophilic center) and is a reagent that forms a bond to its reaction partner (the nucleophile) by accepting both bonding electrons from that reaction partner.

An **electrophilic**（亲电的）center in a molecule is an electron attracting region of positive charge.

In an **elimination reaction**（消除反应）, atoms or groups that are bonded to adjacent atoms are eliminated as a small molecule (e.g., H_2O) and an additional bond is formed between carbon atoms.

In an **electrophilic substitution reaction**（亲电取代反应）, an electrophile replaces another atom or group in a molecule. An example of an electrophilic substitution reaction is the replacement of an H atom in benzene with a nitro (NO_2) group.

An **ester**（酯）is the product of the elimination of H_2O from between acid and an alcohol molecule. Ester have the general formula RCOOR′.

An **ether**（醚）has the general formula ROR′.

The **E, Z system**（E, Z 系统）is a system of nomenclature used to describe the manner in which substituent groups are attached at a carbon-to-carbon double bond.

Fats（脂肪）are triglycerides in which saturated fatty acid components predominate.

A **Fischer projection**（费歇尔投影式）is a two-dimensional representation of a three-dimensional structural formula. It shows how the stereochemistry at a chiral carbon atom is represented in two dimensions, and how the carbon-chain backbone is arranged on the page.

Free radicals（自由基）are highly reactive molecular fragments containing unpaired electrons.

A **functional group**（官能团）is an atom or grouping of atoms attached to a hydrocarbon residue, R. The functional group often confers specific properties to an organic molecule.

Geometric isomerism（几何异构）in organic compounds refers to the existence of nonequivalent structures (cis and trans) that differ in the positioning of substituent groups relative to a double bond. In complexes the nonequivalent structures are based on the positions at which ligands are attached to the metal center.

Heterocyclic（杂环的）compounds are based on hydrocarbon ring structures in which one or more C atoms is replaced by atoms such as N, O, or S.

A **homologous series**（同源系列）is a group of compounds that differ in composition by some constant unit (CH_2 in the case of alkane).

A **hydrocarbon**（碳氢化合物）is a compound containing the two elements carbon and hydrogen. The C atoms are arranged in straight or branched chains or ring structures.

In a **hydrogenation reaction**（氢化反应）, H atoms are added to multiple bonds between carbon atoms, converting carbon-to-carbon double bonds to single bonds and carbon-to-carbon triple bonds to double or single bonds. It is a reaction, for example, that converts an unsaturated to a saturated fatty acid.

The **hydroxyl group**（羟基）is —OH and is usually found attached to a straight or branched hydrocarbon chain (an alcohol) or a ring structure (a phenol).

A **ketone**（酮）has the general formula $RC(O)R'$.

The **leaving group**（离去基团）is the species expelled from an electrophilic molecule following attack by a nucleophile.

Levorotatory（左旋的）means the ability to rotate the plane of polarized light to the left, designate (—).

Lipids（脂类）include a variety of naturally occurring substances (e.g., fats and oils) sharing the property of solubility in solvents of low polarity [such as in $CHCl_3$, CCl_4, C_6H_6, and $(C_2H_5)_2O$]. In a **liquid**, atoms or molecules are in close proximity (although generally not as close as in a solid). A liquid occupies a definite volume, but has the ability to flow and assume the shape of its container.

A **meta (*m*-) isomer**（间位异构体）has two substituents on a benzene ring separated by one C atom.

A **nucleophile**（亲核试剂）is a reactant that seeks out a center of positive charge as a point of attack in a chemical reaction.

Nucleophilicity（亲核性）is a measure of how readily (how fast) a nucleophile attacks an electrophilic carbon atom bearing a leaving group.

A **nucleophilic substitution reaction**（亲核取代反应）is a reaction between a nucleophile and an electrophile. The nucleophile and an electrophile. The nucleophile attacks at a positive center on the electrophile, and the leaving group is ejected from another point.

Olefins（烯烃）are organic compounds that contain one or more carbon-to-carbon double bonds.

Optical isomerism（旋光异构）results from the presence of a chiral atom in a structure, leading to a pair of optical isomers that differ only in the directing that they rotate the plane of polarized light.

An **organic compound**（有机化合物）is made up of carbon and hydrogen or carbon, hydrogen and a small number of other elements, such as oxygen, nitrogen, and sulfur.

An **ortho (*o*-) isomer**（邻位异构体）has two substituents attached to adjacent C atoms in a benzene ring.

A **para (*p-*) isomer**（对位异构体）has two substituents located opposite to one another on a benzene ring.

A **phenol**（苯酚）has the functional group —OH as part of an aromatic hydrocarbon structure.

A **phenyl group**（苯基）is a benzene ring from which one H atom has been removed: $-C_6H_5$.

Position isomers（位置异构体）differ in the position on a hydrocarbon chain or ring where a functional group (s) is attached.

A **primary carbon**（一级碳原子）is attached to one other carbon atom.

In a **protic solvent**（质子溶剂）, the molecules have hydrogen atoms bonded to electronegative atoms, such as oxygen or nitrogen.

A **quaternary carbon**（四级碳）is attached to four carbon atoms.

A **racemic mixture**（外消旋混合物）is a mixture containing equal amounts of the enantiomers of an optically active substance.

The **R，S system**（R，S 系统）is used to indicate the arrangement of the four groups bonded to a chiral center and to provide names that distinguish between optical isomers.

Saturated hydrocarbon（饱和烃）molecules contain only single bonds between carbon atoms.

A **secondary carbon**（二级碳原子）is attached to two other carbon atoms.

Skeletal isomerism（骨架异构）results from differences in the skeletal structures of molecule having the same position.

S_N1（单分子亲核取代反应）is the designation for a nucleophilic substitution reaction in which the rate determining step is unimolecular.

S_N2（双分子亲核取代反应）is the designation for a nucleophilic substitution reaction in which the rate determining step is bimolecular.

Step-reaction polymerization（逐步聚合反应）is a type of polymerization in which monomers are joined together by the elimination of small molecules between them. For example, a H_2O molecule might be eliminated by the reaction of a H atom from one monomer with an —OH group from another.

A **skeletal structure**（骨架结构）is an arrangement of atoms in a Lewis structure to correspond to the actual arrangement found by experiment.

In **stereoisomers**（立体异构体）the number and types of atoms and bonds in molecules are the same, but certain atoms are oriented differently in space. Cis and trans isomerism is one type of stereoisomerism; optical isomerism is another.

A **stereocenter**（立体中心）is an asymmetric carbon atom.

A **structural formula**（结构式）for a compound indicates which atoms in a molecule are bonded together, and whether by single, double, or triple bond.

Substitution reaction（取代反应）are typical of those involving alkane and aromatic hydrocarbons. In such a reaction a functional group replaces an H atom on a chain or ring.

A **tertiary carbon**（三级碳原子）is attached to three other carbon atoms.

The **torsional energy**（扭转能）is the energy difference between the eclipsed and staggered forms of ethane.

Unsaturated hydrocarbon（不饱和烃） molecules contain one or more carbon-to-carbon multiple bonds.

(19) Biochemistry

Absolute configuration（绝对构型） refers to the spatial arrangement of the groups attached to a chiral carbon atom. The two possibilities and D and L.

Adenosine diphosphate（**ADP**）（二磷酸腺苷） and **adenosine triphosphate**（**ATP**）（三磷酸腺苷） are agents involved in energy transfers during metabolism. The hydrolysis of ATP produces ADP, the ion HPO_4^{2-}, and a release of energy.

An **α-amino acid**（α-氨基酸） is a carboxylic acid that has an amino group（—NH_2） attached to the carbon atom adjacent to the carboxyl group（—COOH）.

A **carbohydrate**（碳水化合物） is a polyhydroxy aldehyde, a polyhydroxy ketone, a derivative of these, or a substance that yields them upon hydrolysis. Carbohydrates can be viewed as "hydrates" of carbon, in the sense that their general formulas are $C_x(H_2O)_y$.

The **cell**（细胞） is the fundamental unit of living organisms.

Denaturation（变性） refers to the loss of biological activity of a protein brought about by changes in its secondary and tertiary structures.

Deoxyribonucleic acid（**DNA**）（脱氧核糖核酸） is the substance that makes up the genes of the chromosomes in the nuclei of cells.

An **enzyme**（酶） is a high molar mass protein that catalyzes biological reactions.

Isoelectric point（等电离点）, pI, of an amino acid is the pH at which the dipolar structure or "zwitterion" predominates.

Metabolism（代谢） refers to the totality of the chemical reactions occurring in living organisms.

A **monosaccharide**（单糖） is a single, simple molecule having the structural features of a carbohydrate. It can also be called a simple sugar.

Nucleic acids（核酸） are cell components comprised of purine and pyrimidine bases, pentose sugars, and phosphoric acid.

Oils（油） are triglycerides in which unsaturated fatty acid components predominate.

Oligosaccharide（寡糖；低聚糖） are carbohydrates consisting of two to ten monosaccharide units.

A **peptide bond**（肽键） is formed by the elimination of a water molecule from between two amino acid molecules. The H atom comes from the —NH_3 group of one amino acid and the —OH group, from the —COOH group of the other acid.

In a **polypeptide**（多肽）, a large number of amino acid units join together through peptide bonds.

Polysaccharide（多醣，多聚糖） is a carbohydrate (such as sugar or cellulose) consisting of more than ten monosaccharide units.

Primary structure（一级结构） refers to the sequence of amino acid in the polypeptide chains that make up a protein.

A **protein**（蛋白质） is a large polypeptide, that is, having a molecular mass of 10000 u or more.

Quaternary structure（四级结构）is the highest order structure that is found in some proteins. It describes how separate polypeptide chains may be assembled into a large, more complex structure.

A **reducing sugar**（还原糖）is one that is able to reduce Cu^{2+} (aq) to red, insoluble Cu_2O. The sugar must have available an aldehyde group, which is oxidized to an acid.

Ribonucleic acid (RNA)（核糖核酸）, through its messenger RNA (mRNA) and transfer RNA (tRNA) forms, is involved in the synthesis of proteins.

The **secondary structure**（二级结构）of a protein describes the structure or shape of a polypeptide chain, for example, a coiled helix.

Saponification（皂化反应）is the hydrolysis of a triglyceride by a strong base. The products are glycerol and the soap.

Soaps（肥皂）are the salts of fatty acids, e.g., RCOONa, where the R group is a hydrocarbon.

A **sugar**（糖）is a monosaccharide (simple sugar), a disaccharide, or an oligosaccharide containing up to ten monosaccharide units.

The **tertiary structure**（三级结构）of a protein refers to its three-dimensional structure, for example, the twisting and folding of coils.

Triglycerides（甘油三酯）are esters of glycerol (1,2,3-propanetriol) with long-chain monocarboxylic (fatty) acids.

A **zwitterion**（两性离子）is a compound (for example, an amino acid or polypeptide) containing both acid and base groups. Zwitterions, at neutral pH, typically have simultaneously positively charged groups (cations) and negatively charge groups (anions).

Appendix C Some Laboratory Apparatus

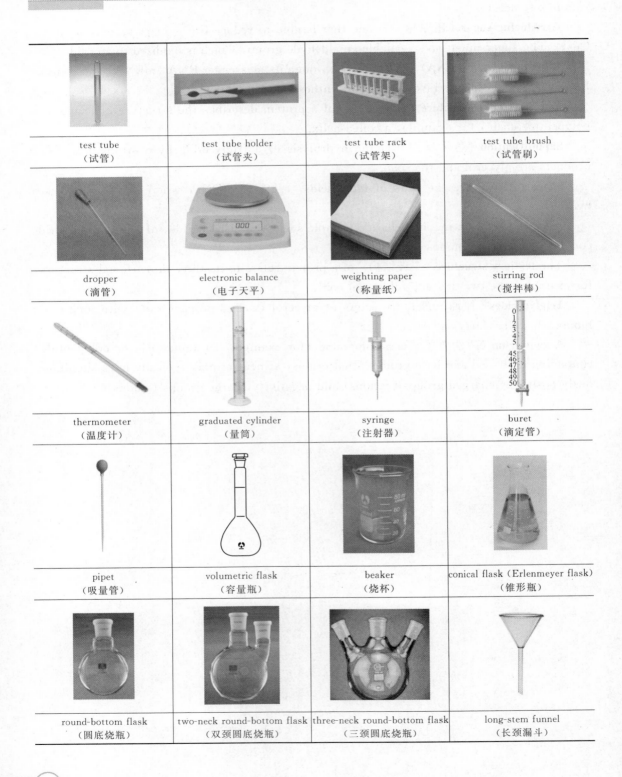

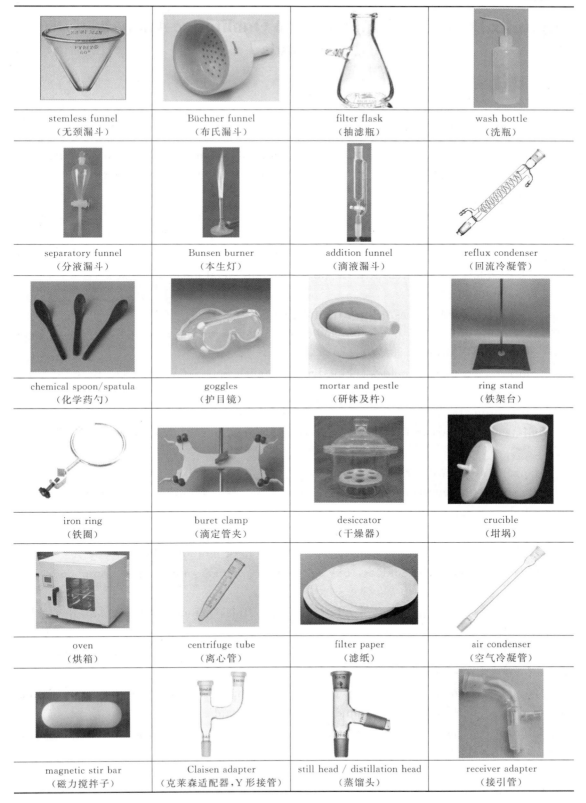

Appendix D Simple Guidelines to Qualitative Analysis of Inorganic Compounds

The following notes are intended to serve as a rough guideline in assisting with the relation between certain observations and the chemical nature of a limited number of compounds. From your knowledge of the periodic table and general chemistry, students should be able to make sensible extrapolations to other compounds not included here.

(1) Color

Colored substances include many metallic oxides and sulfides, transition metal compounds [e.g. Cu(Ⅱ) blue, Ni(Ⅱ) green, Fe(Ⅲ) yellow, Cr(Ⅲ) deep green or purple] and elements such as sulfur (yellow) and iodine (purple).

Lack of color is associated with the non-transition metal ions and most anions.

(2) Solubility in Water

Water soluble substances include the following:

Compounds of sodium, potassium and the ammonium ion, all nitrates(Ⅴ) (except basic nitrates) and most nitrates(Ⅲ).

Chlorides, bromides, iodides, chlorates(Ⅴ) and bromates(Ⅴ) (except those of silver (Ⅰ), mercury(Ⅰ), copper(Ⅰ) and basic salts). Iodates(Ⅴ) of alkali metals. [Lead halides and barium bromate(Ⅴ) are sparingly soluble in cold water.]

Sulfates(Ⅴ), except those of barium, strontium, lead, calcium, mercury(Ⅱ) and silver(Ⅰ).

Manganates(Ⅶ), chlorates(Ⅰ).

Ethanoates and methanoates. [Siver(Ⅰ) and mercury(Ⅰ) ethanoates and methanoates and lead(Ⅰ) methanoate are sparingly soluble.]

(3) Action of Heat on the Solid Sample

Hydrates give water vapor which condenses at top of ignition tube.

Some hydrates, higher oxides, oxy-salts change color, some organic compounds char (go black).

Ammonium salts, mercury(Ⅱ) halides and sulfide, arsenic(Ⅲ) and antimony(Ⅱ) oxides, sulfur and iodine sublime.

Nitrates(Ⅲ) and nitrates(Ⅴ), except those of potassium, sodium and ammonium, give brown fumes of nitrogen dioxide.

Many carbonates, all hydrogen carbonates, ethanedioate (these also give carbon monoxide) and some organic compounds give carbon dioxide.

Sulfates(Ⅳ) (except those of sodium and potassium), thiosulfates(Ⅴ) and some sulfates(Ⅴ) give sulfur dioxide.

Some sulfides give hydrogen sulfide.

Many ammonium salts give ammonia in the course of dissociation.

Methanoates give hydrogen (and other gases), other organic compounds may give flammable vapors with characteristic odors e. g. propanone from ethanoates, methanal from methanoates.

Some higher oxides and oxy-salts evolve oxygen.

(4) Flame Test

Sodium compounds give a golden yellow color.

Potassium compounds give a lilac color appearing crimson when viewed through blue glass.

Calcium compounds give a yellowish red color.

Strontium and lithium compounds give a crimson color.

Barium and copper compounds give a green color.

Lead, arsenic, antimony and bismuth compounds give a greyish blue color (and attack the platinum wire).

(5) Action of Dilute Hydrochloric and Sulfuric (VI) Acids on the aqueous solution of the sample

Insoluble chlorides maybe precipitated by dilute HCl and insoluble sulfates(VI) by dilute H_2SO_4.

Carbonates and hydrogen carbonates give carbon dioxide.

Nitrates(III) give a blue solution of nitric(III) acid in the cold, and brown fumes of nitrogen dioxide on heating.

Sulfates(IV) and thiosulfates(VI) give sulfur dioxide. Thiosulfates(VI) also give a precipitate of sulfur.

Sulfides give hydrogen sulfide.

Chlorates(I) give chlorine.

(6) Action of Concentrated Sulfuric (VI) Acid on the Solid Sample

Hydrates may change color, organic matter may char (go black).

Halides give the corresponding hydride but in addition, bromides and iodides give some of the corresponding halogen.

Nitrates(V) give on strong heating brown fumes of nitrogen dioxide and acid vapors of nitric(V) acid which fumes in air.

Methanoates give carbon monoxide (even at room temperature), ethanedioates give carbon monoxide and dioxide, ethanoates give a smell of vinegar.

Certain reducing agents give sulfur dioxide [by reduction of the sulfuric(VI) acid].

(7) Action of Concentrated Nitric (V) Acid on the Solid Sample

Reducing agents reduce the nitric (V) acid giving brown fumes of nitrogen dioxide; bromides and iodides give bromine and iodine respectively in addition.

Hydrated higher oxides of metals such as tin and antimony are precipitated.

(8) Action of Sodium Hydroxide Solution on the Aqueous Solution of the Sample

Metallic hydroxides, and some oxides (e.g. Ag_2O and HgO) are precipitated and, if amphoteric, dissolve in excess sodium hydroxide solution [e.g. Zn, Al, Pb and Cr(III)].

Some hydroxides are colorless (e.g. Zn, Al, Pb, Mg), whereas others are colored [e.g.

Cu(II) blue going black on heating; Ni(II) green; Fe(III) rust-red; Co(II) blue going pink on heating and brown on exposure to air; Cr(III) greyish green]

Ammonium salts and amides give ammonia on heating.

(9) Action of Ammonia Solution on the aqueous solution of the sample

Metallic hydroxides and a few oxides are precipitated, some dissolve in ammonia solution to form ammines which may be colorless (e.g. Ag) or colored [e.g. Cu(II) deep blue; Ni(II) blue; Cr(III) violet; Co(II) yellow going red on exposure to air].

(10) Action of 0.2mol/L $KMnO_4$, in Dilute H_2SO_4 on the aqueous solution of the sample

Reducing agents such as iron(II), mercury(I) and tin(II) salts, halides, sulfates (IV), sulfides, methanoates and ethanedioates reduce manganate(VII) to hydrated manganese(IV) oxide or to manganese(II) at room temperature or on heating.

(11) Action of Hydrogen Sulfide Solution on the aqueous solution of the sample

Solutions of some cations give precipitates of sulfides (often highly colored). Those of lead(II), silver(I), mercury(II), copper(I), iron(II), cobalt(II) and nickel(II) are black, bismuth(III) and tin(II) are brown, cadmium(II), tin(IV), and arsenic(III) and (V) are yellow, antimony(III) and (V) are orange, zinc(II) is white and manganese (II) is pale pink.

Many oxidizing agents, e.g. iron(III) are reduced and colloidal sulfur is formed.

(12) Action of Zinc and Dilute Sulfuric (VI) Acid on the aqueous solution of the sample

Higher oxidation states of transition metal ions are reduced to lower oxidation states with different colors, e.g. Fe(III) to Fe(II).

Some metal ions, e.g. those of Pb, Cu, Bi, are reduced and the metals deposited on zinc. Some metal ions, e.g. those of As(III) and (V), Sb(III) and (V), are reduced to their hydrides (smell) and the elements (black powder).

Some anions are reduced e.g. sulfates(IV) to hydrogen sulfide.

(13) Properties of and Tests for Gases

Hydrogen. Colorless, odorless, burns with a pale blue fame and slight explosion.

Carbon dioxide. Colorless, odorless, turns lime water milky.

Sulfur dioxide. Colorless, characteristic odor, turns dichromate(VI) impregnated filter paper green.

Hydrogen sulfide. Colorless, characteristic odor, tums lead(II) ethanoate impregnated filter paper black.

Chlorine. Yellowish green, characteristic odor, bleaches moist litmus paper, turns starch-iodide paper blue.

Oxygen. Colorless, odorless, relights a glowing splint.

Nitrogen dioxide. Brown gas, characteristic odor. Turns starch-iodide paper blue.

Bromine. Reddish brown vapor, characteristic odor. Turns fluorescein impregnated paper red and starch-iodide paper blue.

Iodine. Purple vapor, characteristic odor. Turns starch impregnated filter paper blue.

Ammonia. Colorless, characteristic odor. Turns red litmus blue. Gives white fumes with hydrogen chloride.

Carbon monoxide. Colorless, odorless, burns with a pale blue flame.

Hydrogen chloride and bromide. Colorless, characteristic odors. Give white fumes with ammonia.

Nitrogen. Colorless, odorless, inert.

References

[1] Ebbing D D, Gammon S D. General Chemistry. 11th ed. Cengage Learning, 2017.

[2] Petrucci R H, Herring F G, Madura J D, et al. General Chemistry: Principles and Modern Applications. 11th ed. Pearson Canada, Inc., 2017.

[3] Burdge J, Overby J. Chemistry: Atom First. 3rd ed. McGraw-Hill Education. 2018.

[4] Burrows A, Holman J, Parsons A, et al. Chemistry. 3rd ed. Oxford University Press. 2017.

[5] Christian G D, Dasgupta P K, Schug K A. Analytical Chemistry. 7th ed. John Wiley& Sons, Inc., 2014.

[6] Levine I N. Physical Chemistry. 6th ed. McGraw-Hill Companies, Inc., 2009.

[7] Loudon M, Parise J. Organic Chemistry. 6th ed. W. H. Freeman and Company, 2016.

[8] Karty J M. Organic Chemistry: Principles and Mechanisms. 2nd ed. W. W. Norton & Company, Inc., 2018.

[9] Bursten B E, Murphy C J, Woodward P M, et al. Chemistry: The Central Science. 14th ed. Pearson Education, Inc., 2018.

[10] Skoog D A, West D M, Holler F J, et al. Fundamentals of Analytical Chemistry. 10th ed. Cengage Learning, 2021.

[11] Klein D. Klein's Organic Chemistry. 3rd ed. John Wiley& Sons, Inc., 2018.

[12] Chemistry 2e. https://openstax.org/details/books/chemistry-2e. 2019.

[13] Dodd J S. The ACS Style Guide: A Manual for Authors and Editors. 2nd edition, American Chemical Society, Washington, DC, 1992.

[14] Coghill A M, Garson L R. The ACS Style Guide: Effective Communication of Scientific Information. 3rd edition, American Chemical Society, Washington, DC, 2006.

[15] Svehla G. Vogel's Textbook of Macro and Semimicro Qualitative Inorganic Analysis. 5th edition, Longman, New York, 1979.